Synthesis Lectures on Information Security, Privacy, and Trust

Series Editors

Elisa Bertino ⓘ, Purdue University, West Lafayette, IN, USA

Elena Ferrari, University of Insubria, Como, Italy

The series publishes short books on topics pertaining to all aspects of the theory and practice of information security, privacy, and trust. In addition to the research topics, the series also solicits lectures on legal, policy, social, business, and economic issues addressed to a technical audience of scientists and engineers. Lectures on significant industry developments by leading practitioners are also solicited.

Marco Anisetti · Claudio Agostino Ardagna ·
Ernesto Damiani · Nabil El Ioini

A Journey into Security Certification

From the Cloud to Artificial Intelligence

 Springer

Marco Anisetti
Department of Computer Science
Università degli Studi di Milano
Milan, Italy

Claudio Agostino Ardagna
Department of Computer Science
Università degli Studi di Milano
Milan, Italy

Ernesto Damiani
Secure Cyber-Physical Systems Center
Artificial Intelligence and Intelligent Systems
Institute
Khalifa University
Abu Dhabi, United Arab Emirates

Nabil El Ioini
Faculty of Science and Engineering, School
of Computer Science
University of Nottingham Malaysia
Semenyih, Selangor, Malaysia

Department of Computer Science
Università degli Studi di Milano
Milan, Italy

ISSN 1945-9742   ISSN 1945-9750 (electronic)
Synthesis Lectures on Information Security, Privacy, and Trust
ISBN 978-3-031-59723-7   ISBN 978-3-031-59724-4 (eBook)
https://doi.org/10.1007/978-3-031-59724-4

This Springer imprint is published by the registered company Springer Nature Switzerland AG
The registered company address is: Gewerbestrasse 11, 6330 Cham, Switzerland

If disposing of this product, please recycle the paper.

Preface

Verifying that an IT system holds a certain property and will continue to hold it once in production is a tricky task. It certainly requires some experience with the underlying hardware and software technology, as well as familiarity with assurance, validation, and verification techniques. Knowledge representation expertise is needed to design software certificates that contain all the information necessary for reproducible, independent assessment of the desired properties. A fourth ingredient is the certification process' own auditability: it must be carried out in such a way that it can be audited by some trusted authority that will endorse its results. Another essential ingredient is *context integrity*: the certification process should be carried out in the same context (e.g., in the same execution environment) that the system will encounter after deployment. The recent history of the domain is full of failures, most of them due to lack of context integrity. IT virtualization has made it difficult to predict the actual execution contexts; also, some suppliers have been known to doctor the context at verification time, using *defeat devices* that detect when testing is ongoing and kick in to ensure that tests will certify a property the system will not hold in production. In this book, which comes twenty years after our seminal work on open systems certification [1], we propose an entirely new approach to handling context integrity, combining intelligent verification methods, possibly based on Artificial Intelligence, with automated run-time adaptation of certification framework. We argue that our approach will enable verification of the modern distributed systems underlying AI deployment. In other words, you need intelligence to verify intelligence. Certification requires a balance between the involvement of practitioners, who have the expertise, experience, and interest in the entity to certify, and the involvement of external parties, who have the authority, independence, and impartiality to earn the trust of the users. As the limerick below goes, a trustworthy signature is the key to any credible certification.

```
There once was a programmer named Brode
Who wanted to certify some modular code
He worked so hard
To test every part
But then he forgot to sign what he wrote.
```
[1]

Milan, Italy   Marco Anisetti
Milan, Italy   Claudio Agostino Ardagna
Abu Dhabi, United Arab Emirates   Ernesto Damiani
Semenyih, Malaysia   Nabil El Ioini

Reference

1. E. Damiani, C. A. Ardagna, and N. El Ioini, *Open source systems security certification.* Springer Science & Business Media, 2008.

---

[1] Also GPT tried hard, but could not improve this limerick...

Contents

Introduction

1

Distributed systems are everywhere. They include emerging computation and communication environments like the cloud and the Internet of Things (IoT), novel computational architectures such as the ones based on microservices, as well as new development practices like Dev-Ops. Distributed systems also power the training and deployment of Machine Learning, including Generative Machine Learning models based on Large Language Models (LLMs). These distributed systems delivered unprecedented scalability and are now the fabric of our society, as humans are affected by their operation as users, customers and even as components, every day of our lives. Still, a looming shade may still spoil distributed systems' success: the increasing difficulty of guaranteeing they will behave as we expect. No need to think of "rebel machines": simply, systems composed of millions of distributed hardware and software components have many modes for failing, some of them triggered by intentional attacks or by the dis-functional behavior of their human users.

This work starts from the notion that when distributed systems' operation depends on operating conditions and on (possibly dis-functional) human behavior, a fresh approach is needed to their *security assurance*. According to [1], assurance is *the way to gain justifiable confidence that an infrastructure and/or applications will consistently demonstrate one or more non-functional (e.g., security) properties, and operationally behave as expected despite failures, malfunctioning or attacks.*

Indeed, assurance is a much wider and deeper notion than security. It includes all methodologies for collecting and validating evidence supporting specific non-functional properties. It also includes audit, compliance and certification techniques. Today, almost any type of activity, especially in critical domains like healthcare and transportation, is captured in audit trails, whether automated or manual. When distributed systems are used to handle critical applications, digital assurance is essential to avoid oversights and to control liability.

In this book, we focus on certification as the primary means of delivering assurance. We describe its evolution across all sectors of Information and Communication Technology, and discuss how it can be applied to the new generation of intelligent distributed systems.

1.1   Context and Motivation

Distributed systems offer many benefits, such as resource sharing, scalability, fault tolerance, and transparency. They enable thousands of applications and services we use every day, such as online banking, search engines, social media, online gaming, e-healthcare, and more. Distributed systems permeate all aspects of our lives; however, compared with centralized systems of the past, they may fail or be compromised in ways that are much more difficult to predict. There is little doubt that a new approach to assurance is essential; but how to use evidence of distributed systems' correct operation to support their certification is still an open problem, similar to the one we analyzed in our seminal book on open systems software certification [2] but even more critical. Many users, including companies and organizations, have little trust in the correctness and reliability of the distributed systems and services they use every day. Our past work [2] analyzed the problem of software system certification from a wide perspective, considering issues to be managed when very large software systems (e.g., the Linux operating system) underwent a complete certification process. We presented different certification schemes, but chose to focus on Common Criteria [2], the de-facto standard for software system certification. This book takes a fresh approach and presents the evolution of security certification in the framework of ICT evolution.

In the last decade, industries, as well as governments, have been at the forefront of the effort pushing towards the definition of new certification schemes that address the peculiarities of modern distributed systems. For instance, The European Network and Information Systems Security Agency (ENISA) has been appointed by the European Commission as the responsible for developing EU cyber-security certification, which provides the evidence supporting a given level of trust. The mission of ENISA in the area of the EU cyber-security certification framework is *"to proactively contribute to the emerging EU framework for the ICT certification of products and services and to carry out the drawing up of candidate certification schemes in line with the Cybersecurity Act, and additional services and tasks"*.

In this book, we describe the huge research effort devoted to finding methods for checking distributed system quality, which may result in security, safety, and dependability guarantees. Specifically, researchers are addressing the following questions:

- What can we certify today? How can we address the peculiarities of distributed systems?
- How can we trust a certificate to be accurate across distributed system changes?
- How can we measure our trust on dynamic and evolving systems?
- What can be done at design time to integrate development and certification processes?
- How can we address composite systems?

We focus on the certification of security-related properties, that is, the process of proving a number of high-level security properties by collecting a sufficient amount of evidence. Traditionally, security certification provided customers with a quantitative measure of the security level of ICT products, supporting the selection of distributed systems that best match their security needs. A wide range of certified ICT products exist, including operating systems, firewalls, and smart cards. However, among those certified products just a few are modern systems based on (micro-)services. Furthermore, among existing certification schemes, just a few are applicable to modern systems that: (i) are complex and composite, (ii) subject to quick versioning and adaptation, (iii) mix heterogeneous technologies.

The discussion in this book is driven by five main requirements, that are summarized below.

Multi-dimensional certification. Certification is not only a matter of evaluating software artifacts (e.g., code base, system execution). Certification concerns many aspects of a system life cycle, including the development process, the verification process, and beyond (e.g., the deployment process). A certification scheme should consider evidence collected in multiple dimensions, contributing to build a more correct and comprehensive view of the certified system.

Composite certification. Modern systems are implemented as a smart composition of services and components. Certifying a system as a whole introduces inaccuracy in the certification results. The need emerges for certification schemes that are capable of inferring the properties of the whole system starting from the properties of its components.

Continuous certification. Modern systems are modified and deployed at fast rates, with a dozen of service versions released every day. Modern certification schemes must reduce the monetary costs and the time needed to certify a specific system, supporting quick and partial re-certification across system changes. Smart certificate life cycle management approach must be defined, increasing certificate support over time.

Chain of trust. Traditional certification assumes a trusted third party (i.e., certification authority) that is responsible for the entire certification process. Modern systems are however continuously developed and deployed, and the corresponding certification process lasts for the entire system life cycle. The assumption on the availability of a certification authority for the entire certification process is therefore not sound, and the need of new chains of trust routed at a trusted third party emerges.

Deterministic versus non-deterministic. Modern systems are quickly moving from deterministic to non-deterministic behavior driven by Machine Learning (ML) and Artificial Intelligence (AI). This evolution hampers existing schemes and introduces the need of novel certification approaches that accomplish the peculiarities ML/AI-based systems.

1.2   Structure of the Book

Besides this Introduction, this book contains five Chapters. For the reader's convenience, in this section we briefly summarize their content.

Chapter 2 presents a summary of the history of software security certification. It first presents the evolution of certification schemes for traditional software systems, which resulted in the definition of Common Criteria. Then, it describes a common definition of certification process, used throughout the book.

Chapter 3 presents an overview of evidence-based certification solutions for cloud services. It first presents the building blocks of a certification. It then describes the certification process and its life cycle management. It further discusses how to manage incremental certification, addressing system and contextual changes through certificate adaptation. It finally presents an encompassing certification framework and the corresponding chain of trust.

Chapter 4 presents an overview of certification of modern composite systems including: (i) *multi-factor certification* that goes beyond traditional certification built on service artifacts only; (ii) *composite service certification*; (iii) *certification of the deployment environment*; (iv) *certification of the development process*.

Chapter 5 presents the new trends in modern distributed system certification, focusing on the challenges to be addressed in the certification of cloud-edge distributed systems and machine learning-based systems.

Chapter 6 draws our concluding remarks, outlining some open research directions.

References

1. C. Ardagna, R. Asal, E. Damiani, and Q. Vu, "From Security to Assurance in the Cloud: A Survey," *ACM Computing Surveys*, vol. 48, no. 1, August 2015.
2. E. Damiani, C. A. Ardagna, and N. El Ioini, *Open source systems security certification*. Springer Science & Business Media, 2008.

History of Software Security Certification

Today, computer systems play a central role in all aspects of human life, including in the operation of highly-critical infrastructures, for instance, in the domains of healthcare, transportation, and telecommunication. However, in spite of this pervasive presence of computer systems in our daily activities, many users feel reluctant in trusting them. In the majority of cases, this diffidence can be primarily due to the fact that systems lack a proper evaluation process guaranteeing their reliability and trustworthiness. In this chapter, we propose an overview of software security certification techniques. Then we look at some of the leading certification schemes that have been adopted by governments and the industry over the years.

2.1   Software Security Certification

Software security certification refers to the process of verifying that a software system meets a set of predefined security requirements, following an evidence-based process [1]. This process takes into account different aspects of the software, including, system architecture design, code review, and testing, in order to ensure that some certificate can be awarded to the system.

A variety of certification schemes have been developed over the years, each with its own specific scope, requirements, and standards. Some schemes focus on the technical aspects of the software, such as its performance, security, and reliability, others instead focus on the development process and organizational aspects, such as usage policies and the execution environment. Coming to security, there are many ways to certify the security properties of a software system, depending on the type, scope, and complexity of the system, as well as the standards and regulations that apply to it. Common methods are:

© The Author(s), under exclusive license to Springer Nature Switzerland AG 2025   5
M. Anisetti et al., *A Journey into Security Certification*, Synthesis Lectures on Information Security, Privacy, and Trust, https://doi.org/10.1007/978-3-031-59724-4_2

- Certification exams: These are tests that measure the knowledge and skills of individuals who design, develop, or manage secure software systems. Some popular exams, like CISM and CSSLP, cover topics such as the software lifecycle management, requirements, architecture, design, implementation, testing, deployment, operations, maintenance, and supply chain.
- Security audits: Audits independent assessments that evaluate the security posture and compliance of a software system against a set of criteria. Some examples of security audits are ISO 27001, PCI DSS, and NIST SP 800-53. These methods involve reviewing the documentation, policies, procedures, and controls of the system, as well as conducting interviews, tests, and scans.
- Security testing: Tests are used to identify and exploit vulnerabilities in a software system or its components. Some examples of security testing are penetration testing, vulnerability scanning, and code analysis . Tests can be performed manually or automatically, by internal or external teams, and at different stages of the software development life cycle.

On the organizational side, many different actors are involved in certifying the security of a software system, overseeing the evaluation process by following well-documented processes and protocols [1]. The main task of these organizations is to conduct, on behalf of the user community, the evaluation procedures to ensure that software systems meet the necessary standards. If a software system passes all the evaluation milestones, it is granted a certificate, which is used to prove to potential customers that the software has been independently reviewed and deemed to be of high quality. Obtaining software certificates is beneficial for software suppliers as it increases users' confidence in their products and therefore, their credibility. At the same time, the certificate provides assurance to customers regarding the reliability of their systems and allows them to hold their vendors accountable in case of failures or behaviors that invalidate the issued certificates. In Sect. 2.2 we will review xsecurity certification methods in more detail.

2.2   An Overview of Software Security Certification

The history of software security certification can be traced back to the beginning of the Internet. In the early days of personal computing, there was little need for security certifica-tions, as the ability to seamlessly transfer data from one machine to another was nonexistent. Users were forced to manually move data to the intended recipient. Even though policies and regulations were sometimes put in place to regulate who has access to what and how data are managed, they were at management level rather than at software level. The emergence of the Internet marked a significant turning point. By connecting standalone machines, the transfer of information became faster and more convenient. The Internet bridged the gap, allowing users to rely on software systems to manipulate and transmit data without the need for manual actions, on one side, and reducing transaction costs associated with information exchange, on the other side. The rise of the Internet brought access/data control and privacy issues that challenged the initial view of an open-world wide network capable of providing easy access to resources, while maintaining a high level of security and privacy [2].

The Internet turned out to be a major factor in software market expansion; the software industry exponentially expanded, while software companies targeted different domains, many of them focusing on the enterprise and military segments due to their high financial returns [3]. The growth of the software market eventually came with a major downside, which is an higher risk of provisioning insecure software systems. Governments and agencies took notice of this issue, due to the high impact insecure software systems might have on their operations. As a response, several government agencies introduced ad-hoc security certification schemes for the software market. They aimed to establish rigorous standards and evaluation processes to ensure software products' reliability, integrity, and safety.

2.2.1   Certification Schemes

Generally speaking, a certification process is driven by the *security properties* to be certified (e.g., the so-called *CIA triad*: confidentiality, integrity, authenticity), and carried out collaboratively by three main parties: (i) a *system/service provider* that wants to certify its systems/services; (ii) a *certification authority* managing the overall certification process; and (iii) a *lab* accredited by the certification authority (i.e., *accredited Lab*) that carries out the property evaluation. The earliest software security certification schemes can be traced back to the Seventies, when the U.S. Department of Defense (DoD) established the Trusted Computer System Evaluation Criteria (TCSEC) [4]. TCSEC focused primarily on evaluating the security of computer and software systems by introducing a set of security classes. The TCSEC was later revised and became known as the *Orange Book*, which was a widely used reference for software security certification. In the 1990s, the International Organization for Standardization (ISO) and the International Electrotechnical Commission (IEC) developed the Common Criteria [5], which is a set of internationally-recognized security standards that are still used today for software security certification [1]. The Common Criteria provide a framework for evaluating the security of software systems and other types of information technology products. In addition to the Common Criteria, several other trust models are commonly used for software security certification, including the Capability Maturity Model (CMM), the Payment Card Industry Data Security Standard (PCI DSS), and the Health Insurance Portability and Accountability Act (HIPAA), to name but a few.

In the remainder of this section, we provide more details on software certification schemes, discussing their strengths and weaknesses to the aim of identifying the gaps that need to be addressed by future schemes.

2.2.2 Trusted Computer System Evaluation Criteria (TCSEC)

TCSEC, also known as Orange Book, was the first attempt to come up with a clear and systematic certification process. It has been developed and adopted by the US government for computer systems procurement. TCSEC established a complete evaluation framework for accessing the security features of software systems. The primary objective of TCSEC is to lay the foundations for a common language capable of capturing all the concepts and features subject to the certification process. TCSEC targets four Security Levels (Table 2.1) and focuses on seven main security features (Table 2.2).

TCSEC defined the first systematic process to assess security of IT systems. However, it was specific to the US government and military context and did not meet the needs and expectations of other sectors and countries. Being focused on monolithic software systems, TCSEC failed to adapt to the more composite, modular structure of modern software systems. Additionally, the emergence of new classes of threats, such as the ones linked to network connections, called for updating the standard's evaluation criteria. TCSEC was eventually replaced by the Common Criteria (Sect. 2.2.5), an international standard for evaluating the security of IT products and systems. The Common Criteria is more flexible, comprehensive, and adaptable than TCSEC, as it allows for the definition and evaluation of security requirements according to different domains, environments, and objectives.

2.2.3 Information Technology Security Evaluation Criteria (ITSEC)

Information Technology Security Evaluation Criteria (ITSEC) are a disused European standard for computer security certification, originally developed to assess the security of IT products. The ITSEC were the result of harmonizing security evaluation criteria previously adopted by four countries: France, Germany, the Netherlands and the United Kingdom. The activities were started after noticing the limitations of TCSEC. Merging the contributions of the four countries led to the development of a common evaluation philosophy and software engineering practices. A major goal was defining evaluation criteria equally applicable to military and commercial products and systems. ITSEC 1.0 was first published in May 1990. ITSEC release 1.2, published in June 1991 and endorsed by the Commission of the European Union, was focused to certify IT products used by EU government agencies and large organizations with critical security requirements. ITSEC was in use until the early 2000s, when it was replaced by the Common Criteria for Information Technology Security Evaluation (CC) [6]. ITSEC was divided into three parts:

- *Security Target (ST)*: a document defining the security requirements for a particular IT product.
- *Evaluation Assurance Level*: a measure of the level of assurance that the IT product or system met the requirements specified in the ST.

Table 2.1 TCSEC security levels

Security level	Description	Examples
Level D—Minimal protection	This level includes systems that lack formal security controls. It deals with systems that rely on the security of their operational environment without having a secure implementation themselves	Standalone machines with no connection to the network in a secure environment
Level C—Discretionary security protection	This level includes systems that implement a basic level of security, namely, Discretionary Access Control (DAC). However, the system still relies on the admins to enforce security policies	A typical example is an operating system with user file permissions (e.g., Windows 98)
Level B—Mandatory protection	This level includes systems where there is a clear distinction between admin and regular user roles. Typically, systems implement mandatory access controls (MAC) for users and files, which are used to grant access permissions	Linux operating system that enforces MAC access to system files
Level A1—Verified protection	This level includes systems that undergo formal security verification to show their robustness to complex attacks. This level features rigorous standardisation and auditing processes	A cryptographic module, such as a hardware security module (HSM)
Level A2—Structured protection	This level focuses on structured and systematic security requirement documentation and security policies	An operating system with strong cryptographic and key management controls
Level B1—Labeled security protection	This level extends level B by augmenting the security model with additional labels. The labels define the sensitivity and integrity level associated with each subject and object	VAX/VMS operating systems
Level B2—Structured protection with security policy	This level requires a model of the security policy. This permits both the software product and its policy to be formally verified	Secure Enclave Virtual Memory System (SEVMS)

Table 2.2 TCSEC security features

Feature	Description
Identification and authentication	Identification of all the users and objects in a software system. This could include token, bio-metrics and password-based authentication
Access control	These mechanisms are meant to regulate access control to all systems resources. They include Role-Based Access Control (RBAC), Access Control Lists (ACLs), and Mandatory Access Control (MAC)
Auditing	The ability of the system to maintain tamper-proof security records that can be used to audit the system behavior
Trusted recovery	In case of security breaches, the system can recover while maintaining integrity and its security configurations
Trusted communication	Provide secure communication channels between all resources
Object reuse protection	The ability to clean and flush any secure data before resources can be reused by other users or entities
Accountability	Any activities or actions in the system can be associated with a specific entity or user

- *Security Functional Level*: a measure of the level of security functionality provided by the IT product or system.

Seven assurance levels were defined under the ITSEC scheme, namely:

1. E0: No assurance.
2. E1: Functionally tested.
3. E2: Structurally tested.
4. E3: Methodically tested and checked.
5. E4: Methodically designed, tested, and reviewed.
6. E5: Semi-formally designed and tested.
7. E6: Formally designed and tested.

2.2.4   Canadian Trusted Computer Product Evaluation Criteria (CTEPEC)

Canada published in 1993 its own certification scheme in response to the worldwide race towards establishing rigorous criteria for computer systems' security. CTEPEC is a standard by the Communications Security Establishment (CSE) of Canada. It provides an evaluation criterion for IT products based on a combination of the US TCSEC and the European ITSEC approaches [7]. CTCPEC was instrumental to the creation of the Common Criteria. Table 2.3 presents the six security levels considered by the CTEPEC scheme.

The CTEPEC certification process is carried out by the Canadian Common Criteria Evaluation and Certification Scheme (CCS) laboratories. The process starts when the manufacturer or supplier of a system contacts an accredited lab and submits the necessary documentation depending on the level it is seeking to certify. After a pre-evaluation phase, where all the submitted documentation is reviewed, an evaluation planning phase starts, where the accredited lab outlines the concrete evaluation activities to carry out (e.g., verification methodology, timeline, resource allocation). Next, the evaluation phase starts. This phase executes all the pre-defined activities including testing the product's security features, source code review, vulnerability analysis, and the like. Once this phase is completed an evaluation report is generated, which is then used to decide whether to award the required certification level or not.

To guarantee the validity of the certificate for future releases of the product, a continuous certification maintenance phase has been put in place. This activity requires periodic monitoring of the certified products, to ensure their alignment with the security requirements of the awarded level.

2.2.5   Common Criteria (CC)

The Common Criteria (CC) for Information Technology Security Evaluation is an international standard for computer security certification [1]. It was first developed in the 1990s as a joint effort between the National Security Agency (NSA) in the United States, the Communications-Electronics Security Group (CESG) in the United Kingdom, and the Federal Office for Information Security (BSI) in Germany. CC is recognized and used by many governments and organizations around the world, and it is often used as a basis for government procurement and regulatory compliance, as well as for private sector security certification programs.

The main goal of CC is to present a uniform and consistent security evaluation process that can be used to assess the security of IT products. It primarily targets governments, businesses, and large organizations to the aim of guaranteeing with a high level of confidence that the IT products they use meet a predefined level of security.

The CC standard is composed of three parts:

Table 2.3 CTEPEC levels

Level	Description
E1	Level E1 apply to software systems with non-sensitive data and low-security requirements. Many commercial off-the-shelf (COTS) software products fall into the category including web browsers and software utilities. Security at this level deals with the possibility of the software system affecting the security of the environment it is hosted in
E2	Non-critical and low-risk systems can be certified at this level. These systems are checked for their access controls and authentication/authorisation mechanisms
E3	This level deals with systems that manage sensitive information. To be certified at Level E3, the software system needs to provide evidence that shows stronger security features and stricter controls
E4	This level adds more restrictions on the operating environment of the software system. It mandates that the operational environment needs to guarantee a high level of security including secure communication, secure hardware components and a secure operating system
E5	This level deals with highly classified information. Rigorous evaluation processes are used to verify the security features of the products undergoing the certification process at this level
E6	This is the highest level a software product can achieve under CTCPEC. software systems dealing with highly sensitive data can be certified at this level if they can formally and empirically prove a high level of security for all the considered features

- Protection Profile (PP): a document describing the security requirements for a particular IT product or system. This represents what security means for the organization seeking to certify its IT product.
- Security Target (ST): a document describing how the IT product meets the requirements specified in the PP.
- Evaluation Assurance Level (EAL): a measure of the level of assurance that the IT product meets, according to the requirements specified in the ST.

Over the years, CC has undergone several revisions since it was first introduced. The most recent version, CC:2022,[1] was released in 2022. It supports new technologies and security challenges including IoT and cloud-based systems, as well as the notion of continuous certification.

The CC defines seven Evaluation Assurance Levels (EAL) that represent increasing degrees of assurance.

---

[1] https://commoncriteriaportal.org/cc/index.cfm.

1. EAL1: Functionally Tested.
2. EAL2: Structurally Tested.
3. EAL3: Methodically Tested and Checked.
4. EAL4: Methodically Designed, Tested, and Reviewed.
5. EAL5: Semi-Formally Designed and Tested.
6. EAL6: Semi-Formally Verified Design and Tested.
7. EAL7: Formally Verified Design and Tested.

Common Criteria (CC) provides a common framework and language for specifying, implementing, and evaluating security requirements and features of IT products and systems. It enables mutual recognition and acceptance of security evaluations among different countries and sectors, which reduces costs and increases interoperability. Furthermore CC allows users and customers to choose the IT products and systems that best meet their security needs and expectations, which in turn increases confidence and trust.

Still, CC has some known drawbacks. It is complex and time-consuming to apply and comply with, which requires a lot of resources and expertise from both developers and evaluators. It is not always up-to-date with the latest technologies and threats, which may limit its effectiveness and relevance. Also, CC is not a guarantee of absolute security, as it only evaluates the security claims and features of IT products and systems, not their actual performance or behavior in real-world scenarios.

2.3   Certification Process

Although each of the schemes described so far defines its certification process, they all share the same common pattern. They all rely on an evidence-based certification process that certifies products, practices, and systems. The certification process takes as input the properties and the product/system to be certified and produces as output a machine-readable certificate, containing the evidence that proves the security level of the required properties. The main advantage of an evidence-based certification process is its reliance on empirical evidence to support the claims made about a product/system. A typical evidence-based certification process involves the following steps:

1. Identify the properties to be certified.
2. Identify the requirements to be verified for each property.
3. Gather and review relevant evidence to support the requirements.
4. Evaluate the quality and relevance of the evidence.
5. Determine whether the evidence is sufficient for the required level of security.
6. Award a certificate if the evidence meets the necessary criteria.

Fig. 2.1 Certification process:
conceptual framework

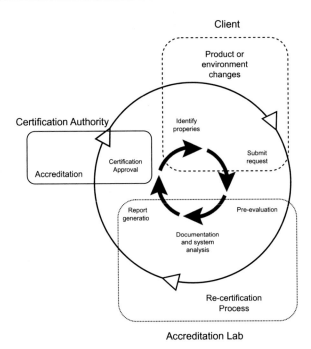

Based on the conceptual framework shown in Fig. 2.1, the certification process is composed of two cycles and is carried out by three actors. The inner cycle is initiated by the client, who defines the scope of the certification by identifying the features to certify and the required certification level. Upon submitting a certification request, the client needs to support the request with all the essential documentation to facilitate the evaluation process.

The second phase involves the accredited lab, which plans the evaluation process and carries out all the involved activities, including product testing, code review, and threat analysis. The accredited Lab generates the evidence needed to certify the product based on the provided documentation and the executed tests, and sends it to the certification authority. If the evidence is sufficient to prove the required property, the certification authority awards a certificate to the client, which includes the certified property, the certification level, and the evidence.

Once the inner cycle is completed, it triggers the outer cycle. The latter monitors the certified product and, in case there are any changes either in the product or its environment that can affect the certified properties, a partial or full re-certification is triggered.

Standard software security certification schemes play a vital role in increasing the trustworthiness of software products. Today, the ISO/IEC 17065 international standard [8] specifies the requirements for all entities that operate product, process, or service certification schemes. It ensures that such entities are competent, impartial, and consistent in their certification activities. Still, certification schemes suffer from several weaknesses. Critical among these weaknesses are the lack of adaptability and the use of a centralized certification approach.

The fast-paced and dynamic nature of the technological landscape, coupled with the ever-evolving classes of threats, poses a considerable challenge for traditional certification schemes. The lack of a dynamic (and fast) way to incorporate up-to-date security practices and new attack vectors can severely limit the effectiveness of the certification process.

Additionally, the software security certification process can greatly benefit from a decentralized approach. Decentralizing the process allows independent actors with different expertise and backgrounds to collaborate and work together in assessing and certifying software security. This decentralized approach can then enhance the robustness and effectiveness of the entire certification process.

The future of software certification is likely to be influenced by several factors, such as the rapid evolution of technologies and threats, which requires software professionals to update their knowledge constantly. Also, certification will have to deal with the growing diversity and complexity of software domains and execution environments.

From the methodology point of view, evidence collection for classic certification schemes has been independent from the development process [9]; today, instead, we expect full integration of certification within agile and DevOps development methodologies, to promote continuous improvement [10].

As we will discuss in the next Sections, an emerging trend is the adoption of cloud-based certification tool kits, which enable automated, flexible software testing. Also, certification is undergoing an alignment of international and industry standards, which enhance the credibility and value of certifications.

References

1. E. Damiani, C. A. Ardagna, and N. El Ioini, *Open source systems security certification*. Springer Science & Business Media, 2008.
2. G. White, E. Fisch, and U. Pooch, "Government-based security standards," *Information Systems Security*, vol. 6, no. 3, pp. 9–19, 1997.
3. M. Campbell-Kelly and D. D. Garcia-Swartz, "The history of the internet: the missing narratives," *Journal of Information Technology*, vol. 28, no. 1, pp. 18–33, 2013.
4. S. B. Lipner, "The birth and death of the orange book," *IEEE Annals of the History of Computing*, vol. 37, no. 2, pp. 19–31, 2015.
5. D. S. Herrmann, *Using the Common Criteria for IT security evaluation*. CRC Press, 2002.
6. M. Gehrke, A. Pfitzmann, and K. Rannenberg, "Information technology security evaluation criteria (itsec)-a contribution to vulnerability?" in *IFIP Congress (2)*. Citeseer, 1992, pp. 579–587.
7. E. M. Bacic, "The canadian trusted computer product evaluation criteria," in *[1990] Proceedings of the Sixth Annual Computer Security Applications Conference*. IEEE, 1990, pp. 188–196.
8. F. Z. Özkan, "A comparative study of iso/iec 17065: 2012 standards, accreditation processes implemented in turkey regarding turkish organic agriculture legislations," *Agric. Sci.*, 2021.

9. A. Colombo, E. Damiani, F. Frati, S. Oltolina, K. Reed, and G. Ruffatti, "The use of a meta-model to support multi-project process measurement," in *Proc. of APSEC 2008*. IEEE, 2008, pp. 503–510.

10. M. Anisetti, C. A. Ardagna, F. Gaudenzi, and E. Damiani, "A continuous certification method-ology for devops," in *Proc. of MEDES 2019*, 2019, pp. 205–212.

Evidence-Based Certification of Cloud Services   3

Cloud-based service certification extends traditional certification schemes to address the peculiarities of dynamic distributed environments. It pursues three main objectives: (i) greater flexibility of the certification process (i.e., models for certification life cycle, target, and process), (ii) adaptability to service evolution and environmental changes (i.e., incremental certification), (iii) soundness of the chain of trust, reducing the involvement of chartered (and costly) Certification Authorities. In this chapter, we first present some evidence-based certification schemes (Sect. 3.1). Then, we introduce the scenario we used throughout the book as our running example (Sect. 3.2), the certification building blocks (Sect. 3.3), and details on how to model a continuous certification process considering a dynamic life cycle and evolving services (Sect. 3.4). We also present the key notion of *incremental certification* to cope with evolving services (Sect. 3.5) and a test-based certification framework (Sect. 3.6). Finally. we present the certificate *chain of trust*, addressing the needs of a trustworthy certification process (Sect. 3.7).

3.1   Evidence-Based Certification Schemes

Evidence-based certification schemes evaluate and certify the performance or quality of an IT system based on empirical data and measurable outcomes. Evidence-based certification schemes for service/based and cloud/based systems consider two main types of evidence:

1. *test-based evidence*, where evidence is collected as the result of testing activities performed by the CA on the target of certification [1–3];
2. *monitor-based evidence*, where evidence is collected in the form of metrics retrieved by a trusted monitoring service, guided by rules defined by the CA [4–12].

© The Author(s), under exclusive license to Springer Nature Switzerland AG 2025   17
M. Anisetti et al., *A Journey into Security Certification*, Synthesis Lectures on Information Security, Privacy, and Trust, https://doi.org/10.1007/978-3-031-59724-4_3

Schemes based on test-based evidence are the most popular ones (e.g., they are used in almost all the evaluation assurance levels of Common Criteria [13] certification scheme). Evidence collection is based on a set of test cases to be executed on the target system. Given the complexity of finding such test cases, some literature focused on automatically generating test cases for certification [14, 15]. These solutions differ mainly in how to use them in the framework of certification [1–3]. Much research and development effort went into the definition of automated test-based evidence collection platform. Among the Research and Innovation (RIA) projects funded by EU on certification, the EU FP7 ASSERT4SOA project [16] pioneered the use of testing evidence to certify Service-Oriented Architecture (SOA) services leading to a complete scheme including the certificate issuing and binding. The EU FP7 CUMULUS project [17] extended the solution in project ASSERT4SOA to cloud computing, including both testing and monitoring evidence.

Certification schemes using monitoring evidence are based on (i) rules expressing conditions that must be satisfied during the monitoring of a given *ToC*, (ii) monitoring assumptions that specify how to record and update variables indicating the state of *ToC* during monitoring. They normally use languages based on Event Calculus [18] for expressing monitoring conditions (e.g., *EC-Assertion+*). This approach is used in the EVEREST monitoring system [19]. Evidence in these schemes is the monitoring events and the period of evaluation must be defined. The evidence verification process must specify the minimum period of monitoring *ToC*, the minimum number of monitoring events, and their representativeness.

Schemes based on Trusted Computing (TC) require that the *ToC* resides on hardware having TPM modules (http://www.trustedcomputinggroup.org/resources/tpm_main_specification) installed. These schemes reach higher trustworthiness thanks to the TPM, but can just support a few properties like software integrity and in most of the cases just at the lowest layers of the cloud stack. Given the nature of TC, the trust chain starts with the trust on the TPM and the physical platform hosting the TPM chip. It is capable of measuring the integrity of the firmware and software in the upper stack reaching the application layer, but in most of the cases requires the support for virtualized TPMs (http://www.trustedcomputinggroup.org/resources/virtualized_trusted_platform_architecture_specification). The assessment scheme of TC-based certification is based on *remote attestation* of *ToC* integrity, which is a functionality provided by trusted computing platforms [20].

TC solutions are also used in the context of certification as a way to verify the integrity of the evidence collected using other schemes like the testing or monitoring ones. Since evidence of different types permits to cover different peculiarities of the system to be certified, Common Criteria as well as FP7 ASSERT4SOA and FP7 CUMULUS, support multiple evidence types including those built on formal proofs. In the case of formal proofs, the evidence is the result of the execution of the proofs on the target system [21]. As in CC, such evidence provides highest trustworthiness, and can be applied just in some specific scenarios and for some properties. Also it requires full access to the system source code to build the formal model. Formal proof-based methods are mostly used for certification of *functional* properties such as correctness [22, 23].

Evidence-based certification schemes have been applied in many relevant scenarios where a trustworthy approach is needed including compliance, verification, and monitoring. More specifically, they support compliance with a given Service/Level Agreements (SLA) [6, 9, 10, 24–27], as a means to implement verification/as/a/service in grid computing [22], behavioral analysis of Network Virtualization Functions [28, 29], as well as monitoring of data integrity in cloud/edge datastores [30].

For the sake of completeness, we need also to mention *non-evidence based* certification schemes. Most of these schemes support the properties of autonomous, highly adaptive systems, and follow these systems' adaptations by updating the properties in the certificates [31–34]. In other words, non-evidence based certification schemes are based on evolving self-declarations, which are modified every time the certified entity changes, in order to reflect a new level of expected performance or quality due to the change. While these non-evidence based certificates may be the only way to provide users with some updated information on the properties of self-adapting systems, it must be underlined that without third-party accreditation, there is no guarantee that a certification scheme can provide any valid assurance, nor that it has a robust audit [35].

3.2   Reference Scenario: Compliance Framework

Our reference scenario is a *compliance framework* used by an accredited lab to collect and verify evidence in the context of a certification process. To this aim, its trustworthiness should be carefully verified and, possibly, certified as discussed in the following of this book. The compliance framework is cloud-native and capable of executing assessment activities on a target system to evaluate the level of compliance with a specific regulation. It has also to be adopted as part of the target life cycle, and therefore included in the target's DevOps pipeline. It is a composition of services implementing specific functionalities needed for compliance evaluation. Given the sensitivity of the activities carried out and data managed, it is mandatory to be certified for some non-functional properties such as confidentiality, integrity, and privacy, to name but a few.

Figure 3.1 shows the architecture of our compliance framework based on (i) compute nodes executing evaluation activities on different hooks (*Execution Manager*) and verifying compliance according to collected artifacts (*Compliance Manager*), (ii) storage nodes storing the compliance controls (*Compliance Control Repository*) and the results of compliance verification (*Artifacts Storage*). We note that the compliance engine implementing this architecture is part of the application life cycle as a first-class citizen, being able to scale and migrate with the application. It exposes APIs that can be called during the different stages of DevOps of the system willing to show compliance.

The *Execution Manager* is in charge of the compliance controls. It must be included in the automatic application deployment procedure ensuring that the hooks of the target application are always reachable. This includes replicating and migrating the Execution Manager

Fig. 3.1 The architecture of
our reference scenario

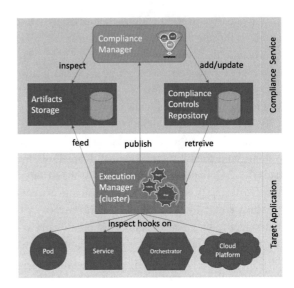

together with the application and elastically scaling based on the observed compliance load
in terms of the number of controls to be simultaneously executed.

The *Compliance Control Repository* includes a set of versioned compliance controls
(e.g., scripts) that can be executed by the Execution Manager. Each control is identified by a
unique name, a version, and requires a number of parameters for setting up the verification.
Its output includes a set of artifacts (e.g., relevant chunks of configuration data) represented
in a machine-readable format. For instance, a control can verify whether a banning service
connected to an access control system works correctly. To do so, the control can access
multiple times, on behalf of a user, a resource without holding the access privileges. If the
corresponding user is banned the control returns a positive result, otherwise a negative result.

Hooks define how compliance controls connect to the target of evaluation and the creden-
tials needed to access it. The *Artifacts Storage* collects the results of the compliance control
execution. The *Compliance Manager* is responsible for verifying compliance by evaluating
the collected evidence. It is also responsible for adding/updating the compliance controls in
the Compliance Control Repository. It is connected to the Execution Manager for run-time
reconfiguration.

3.3   Certification Building Blocks

We now describe the building blocks of a certification process, introducing the notion of
(i) non-functional properties as the scope of the certification scheme, (ii) non-functional
mechanisms as the software components supporting a given non-functional property, (iii)

Target of Certification as the perimeter of the certification in terms of the set of mechanisms to be verified, (iv) evidence as the concrete artifacts of a certification process.

3.3.1 Non-functional Properties

Assessing if a given non-functional property holds for a given system is the final purpose of every certification scheme. The notion of non-functional property is quite well understood in the research community [36–40] and largely adopted for building domain-specific vocabularies for regulations (e.g., [41]), standards (e.g., [42]) and cloud security specifications (e.g., [43]). In the following, we provide an informal definition of some non-functional properties.

- *Confidentiality* is the ability of limiting information access and disclosure to authorized clients only.
- *Integrity* is the ability of preserving the structure and content of information resources.
- *Authenticity* is the ability of ensuring that the clients or objects are genuine.
- *Non-repudiation* is the ability of preventing denial of having sourced an information resource.
- *Robustness* is the ability of a computer system to cope with errors during execution or the ability of an algorithm to continue to operate despite abnormalities in input, calculations, and the like.
- *Availability* is the ability of guaranteeing continuous access to data and resources by authorized clients.
- *Performance* is the ability of guaranteeing specific performance like response time or data ingestion latency.

Since the above properties represent generic security requirements and provide no information on how to achieve them we deemed them as "abstract" non-functional properties. Abstract non-functional property are extended with attributes to define what we call a "concrete" property, as follows.

Definition 3.3.1 (*Non-functional Property*) A non-functional property p can be defined as a pair $(\hat{p}, A_p)$, where $\hat{p}$ is an abstract property and A_p is a set of specific property attributes.

Attributes can include information on the non-functional mechanisms guaranteeing $\hat{p}$ (e.g., access control), a level modeling property strength (e.g., low, medium and high similarly to the CVSS score), and contextual attributes (e.g., *uptime* = 98% for property availability).

Non-functional properties can be organized into a hierarchy $\mathcal{H}_P = (\mathcal{P}, \preceq_P)$ [15], where $\mathcal{P}$ is the set of properties and $\preceq_P$ is a partial order relationship over $\mathcal{P}$. Given two properties

$p_i, p_j \in \mathcal{P}, p_i$ is weaker than p_j (denoted $p_i \preceq_P p_j$) if $p_i.\hat{p} = p_j.\hat{p}$, and $\forall a_k \in A_p$, either the value $v_i(a_k)$ of attribute a_k is not specified or $v_i(a_k) \preceq_{a_k} v_j(a_k)$. We note that expert users can define total order relations $\preceq_{a_k}$ between contextual attributes $a_k \in A_p$.

3.3.2   Non-functional Mechanisms

A non-functional mechanism θ is a software component that supports a given (set of) non-functional properties (e.g., encryption mechanism supporting integrity). It can be formalized as a pair $(\hat{\theta}, A_m)$, where $\hat{\theta}$ is a mechanism type (e.g., digital signature), and A_m is a set of attributes specifying configurations affecting the mechanism behavior (e.g., $standard = ECDSA$). Non-functional mechanisms can be organized in terms of their type $\hat{\theta}$ into a hierarchy $\mathcal{H}_{M_i}$. This hierarchy $\mathcal{H}_{M_i}$ is defined as a pair $(\mathcal{M}_i, \preceq_{M_i})$, where $\mathcal{M}_i$ is the set of mechanisms of the same type, and $\preceq_{M_i}$ is a partial order relationship over $\mathcal{M}_i$ normally defined by domain experts.

For instance, given two mechanisms $\theta_j, \theta_k, \theta_j$ is weaker than θ_k (denoted $\theta_j \preceq_M i\theta_k$) if $\theta_j.\hat{\theta} = \theta_k.\hat{\theta}$, and $\forall a_t \in A_m$, either the value $v_j(a_t)$ of attribute a_t is not specified or $v_j(a_t) \preceq_{a_t} v_k(a_t)$ where the total order relations $\preceq_{a_t}$ between attributes $a_t \in A_m$ are normally defined by expert users. For completeness, a special hierarchy $\mathcal{H}_{M_f} = (\mathcal{M}_f, =)$ where $\mathcal{M}_f = \{(Functional, \{\})\}$ is also defined to model the functional mechanisms of a given service. Both functional and non-functional mechanisms can be logically seen as part of a common hierarchy $\mathcal{H}_M$ of mechanisms $\mathcal{M}$ with a common ancestor denoted as *any*.

Each mechanism is also annotated with a set of events ($\{events\}$) or configurations impacting its execution.

3.3.3   Target of Certification

The *Target of Certification (ToC)*, defines the mechanisms that are part of the certification perimeter for which a given (set-of) non-functional property applies. We note that, in many situations, to support a given non-functional property one or more cooperating mechanisms in the *ToC* are needed.

ToC can be formally defined as (Θ, b), where $\Theta = \{\theta_i\}$ is a set of mechanisms $\theta_i \in \mathcal{M}$ and b specifies the provisioning layer for the system to be certified (i.e., service, platform, infrastructure).

For instance, let us consider a *ToC* with a binding defined at service layer (i.e., $b = $ `<service>`) to be certified for security property $p=$(*Confidentiality*, {`ctx` $= in\text{-}transit/at\text{-}rest$}). The *ToC* includes two mechanisms θ_1 and θ_2 supporting p. Mechanism $\theta_1 = $ (*encryption*, {`algo` $= XML\text{-}encryption$, `protocol` $= WS\text{-}Security$, `level` $= message\text{-}in\text{-}transit$}) refers to a mechanism implementing an encrypted communication channel, mechanism $\theta_2 = $ (*encryption*, {system $= encrypted\ DBMS$}) identifies a mechanism implementing an encrypted DBMS.

3.3.4 Evidence

Evidence *ev* includes the artifacts collected by the certification process to verify whether a given non-functional property is supported or not. Each mechanism θ in the *ToC* can be further modeled in terms of its behavior for evidence collection. Solutions based on Symbolic Transition Systems (STS) [15] have been proposed in literature; they provide an accurate model to the accredited labs for the generation and execution of evidence-collection activities. Evidence supporting the certification process can be of different types (see Sect. 3.1). Being the most largely used, in the following, we concentrate on test-based evidence only.

We can consider three main *categories* of test cases, each of them further specified in terms of *test types*, which specify the test cases needed to support the evaluation of security properties on a given service. More specifically:

- Category functionality: including functional test cases based on valid input. It distinguishes between three test types: (i) *input partitioning* where testing activities are based on different input partitioning strategies (i.e., *Random Input, Boundary Value, Equivalence Partitioning*, and *Fuzzy/Mutation*); (ii) *model control flow* where testing activities are based on path analysis performed on the service model; (iii) *code walkthrough* where testing activities are based on manual analysis of the source code by experts.
- Category robustness: including test cases based on invalid and malformed input. It also includes stress and load tests.
- Category penetration: including test cases based on well-known security attacks. It considers test types in the form of attacker capabilities (e.g., add, modify, copy).

We note that test categories functionality and robustness are needed to check the functional correctness and robustness of a given mechanism θ. Test categories and types, according to [15] can be organized into hierarchies meaning that, assuming the same certified property and test category, a certificate with a given test type can be "stronger" than a certificate with another test type.

Test-based evidence *ev* can be formally defined as follows.

Definition 3.3.2 (*Test-based Evidence*) Let *cat*(.) be a test category, *type*(.) a test type, *ta*(.) a set of test attributes, *tc*(.) a set of test cases, and *tr*(. the results of test case execution. An evidence, denoted ev_i, is a set of 5-tuple $\langle cat(ev_i), type(ev_i), ta(ev_i), tc(ev_i), tr(ev_i) \rangle$.

A test case $tc(.)$ can be further specified in terms of preconditions (Pr), inputs (I), expected outputs (EO), and post-conditions (Po). It is normally designed for a particular objective, such as to exercise a particular execution path or to verify compliance with a specific requirement. A test case $tc(.)$ can be formally defined as follows.

Definition 3.3.3 (*Test case*) A test case $tc(.)$ is a 5-tuple $\langle \theta, Pr, HC, \{(I_1, EO_1), \ldots, (I_n, EO_n)\}, Po\rangle$, where θ is the mechanism under evaluation, Pr is a set of pre-conditions, HC a set of hidden communications, $\{(I_1, EO_1), \ldots, (I_n, EO_n)\}$ a set of pairs (input,expected output), and Po a set of post-conditions.

Test cases can be specified for the following purposes:

1. *Operation-oriented test cases* (OO). They test each single service operation. Each test case $tc(.)$ includes a pair (I, EO), with I the input and EO the expected output related to a specific service operation.
2. *Workflow-oriented test cases* (WO). They test multiple operations in a workflow fashion. Each test case $tc(.)$ defines a specific path in the workflow to be tested and is composed of a set of pairs (I_{o_j}, EO_{o_j}), with o_j the j-th operation in the workflow. We note that, when the output of a given operation is not relevant to the testing activity, $EO_{o_j} = null$.
3. *Implementation-oriented test cases* (IO). They test internal states of a given operation endpoint. This approach requires full access to the source code or its detailed modeling. From a practical point of view, they are similar to the workflow-oriented test cases, each test case is composed of a set of pairs (I_{o_j}, EO_{o_j}). In this case, however, the service implementation is also considered to generate the test cases and to test specific flows to cover internal implementation-specific branches.

3.4   Certification Modeling

Certification schemes for cloud services rely on three modeling layers:

- **Mechanism Model**: it describes how to model the mechanisms θ in *ToC* including the visibility level from white to black box. It is fundamental to drive the evidence-collection process. The more the model is accurate, the better the evidence-collection process can be instrumented. Model quality metrics can be defined with the scope of understanding the level of detail available to carry out our certification.
- **Certification Life Cycle Model**: It describes how to model the evolution of a certificate across time, from its issuing to possible expiration or revocation. The *CA* is traditionally responsible for making decisions on the certificate life cycle, such as, certification issuing, suspension, revocation, or expiration, due to new threats, regulations or auditing activities.

- **Certification Process Model**: It describes how to model the entire certification process including all the building blocks in Sect. 3.3. The certification process is responsible for all evaluation activities needed to produce a certificate for the *ToC* (*issuing phase*), as well as to continuously verify the validity of the certificate against contextual changes, to the aim of reducing unnecessary certificate revocation and re-certification [33] (*post-issuing phase*). It specifies how the evidence is collected and evaluated. Evidence quality metrics can be specified to evaluate the strength and trustworthiness of the certificate.

Software mechanisms modeling is largely covered in literature in the context of software testing and assessment. In the following we (i) assume that the security mechanisms are modeled in a way suitable for testing, such as using STS [15], and (ii) concentrate on a certification life cycle and process suitable for testing evidence.

3.4.1 Certificate Life Cycle Model

Cloud-related certificates must evolve in time following the evolution of the target service, threats and regulations. This evolution is much more dynamic than for traditional software and has to be automated as much as possible. For this reason, a certificate life cycle l is defined; it has the scope of modeling the certificate evolution from its issuing to possible expiration or revocation. It is handled (i) offline by the *CA* in a static and asynchronous way as a reaction to new threats and regulation changes and (ii) online and at run-time on the basis of the evidence collected by the accredited lab, since the static intervention of a *CA* is not always feasible, possible.

Definition 3.4.1 (*Life cycle*) Life cycle l is a deterministic finite state automaton $G^l(V^l, E^l)$ where each vertex $v \in V^l$ represents the certificate state (e.g., not-issued, issued, suspended, revoked) using a label *label*(v) and each edge $e = (v_i, v_j) \in E^l$ represents the transition between two states. We note that each transition e is labelled with a condition *cond*$_e$ over certificate evidence that regulates its execution.

Figure 3.2 shows an example of the life cycle automaton with transition conditions. For instance, edge $e_{NI,I}$ is labeled with a condition *cond*$_{e_{NI,I}}$ stating that at least 95% of the test cases $tc(.)$ need to be successfully executed (i.e., *cond*$_{e_{NI,I}} = 0.95tc$) for the certificate to move from state NI to state I.

3.4.2 Certification Process Model

We first present our conceptual framework supporting the certification process and then the model of our certification process incarnating our conceptual framework.

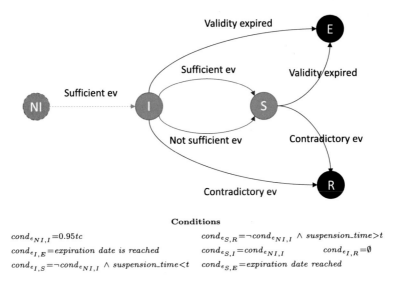

Conditions

$cond_{e_{NI,I}}=0.95tc$ $cond_{e_{S,R}}=\neg cond_{e_{NI,I}} \wedge suspension\_time>t$

$cond_{e_{I,E}}=expiration\ date\ is\ reached$ $cond_{e_{S,I}}=cond_{e_{NI,I}}$ $cond_{e_{I,R}}=\emptyset$

$cond_{e_{I,S}}=\neg cond_{e_{NI,I}} \wedge suspension\_time<t$ $cond_{e_{S,E}}=expiration\ date\ reached$

Fig. 3.2 Life cycle with states Not Issued (NI), Issued (I), Suspended (S), Expired (E), Revoked (R) and examples of conditions on transitions

3.4.2.1 Conceptual Framework

We consider three models driving the certification process: (i) the abstract declarative model defined by the CA describing the property p to be certified, the target of the certification activities ToC and evidence to be collected, and the evidence collection activities to be executed (Certification Model (CM) Template), (ii) the procedural model of the concrete activities to be carried out by the accredited lab on the given implementation of the target (Certification Model (CM) Instance), (iii) the model of the certificate to be released. Figure 3.3 shows the conceptual framework at the basis of the certification process.

The aim of a certification process is to prove a property p for a given ToC. It is driven by a CM Instance. CM Instance is generated as a concrete incarnation of a given CM Template and verified against it to ensure its validity. The evidence is continuously collected, according to the CM Instance collection procedure. In case the evidence is not sufficient to prove a given non-functional property (dashed arrow in Fig. 3.3) the corresponding certificate cannot be awarded.

CM Template T is defined by the CA and drives the entire certification process. It represents the cornerstone of the certification chain of trust, which is grounded on the correctness of the certification template itself, as it is designed and signed by the CA (see Sect. 3.7). It is a declarative high-level representation of the process including requirements specification, configurations, and activities for the certification of a property for a given (class of) ToC.

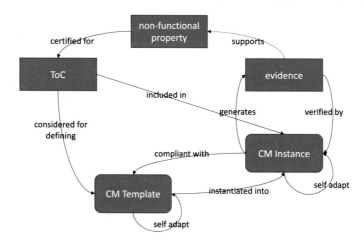

Fig. 3.3 The conceptual framework

Definition 3.4.2 (*CM Template T*) A CM Template T is defined as 5-tuple of the form $\langle p, ToC, m, ev, l\rangle$ and signed by the CA. The requirements on the *ToC* are expressed in the CM Template T in terms of the mechanisms to be used to support a given property p. The certification activities are expressed using a model m describing the evidence collection, the evidence ev to be collected, and the certificate life cycle l.

The evidence collection model m describes the set of execution flows $\Phi(m)$ targeting the mechanism in Θ, which must be evaluated to certify p. The flows are structured to cover the different relevant purposes of testing (i.e., operation OO, workflows WO and Implementation IO, see Sect. 3.3.4). Evidence collection flows $\Phi(m)$ are associated with evidence ev to be produced. The evidence ev in T is expressed in an abstract format defining, for instance, the inputs I_n and expected outputs EO_n in the form of partitions of the corresponding input and output domains (see Definition 3.3.3). The life cycle l is also part of the CM Template. It is defined by the CA to describe the certificate evolution across time depending on the outcomes of the evidence collection and other external events like expiration and revocations. We note that at CM Template level, the mechanisms $\theta \in ToC$ can be defined at a very abstract level just including the type only (i.e., $(\hat{\theta}, \emptyset)$) or at a more concrete level including specific attributes (i.e., $(\hat{\theta}, A_m)$).

Example 3.4.1 (*T*) Let us consider property *confidentiality* and the *Artifacts Storage* service of the compliance framework in our reference scenario (Sect. 3.2). A CM Template T includes the security property $p = (Confidentiality, \{\texttt{ctx} = in\text{-}transit/at\text{-}rest\})$ and target of certification $ToC = (\{\theta_1, \theta_2\}, <service>)$, where $\theta_1 = (encryption, \{\texttt{level} = internal\text{-}communications\})$ is an encryption mechanism securing internal service communications, $\theta_2 = (encryption, \{\texttt{level} = data\text{-}at\text{-}rest\})$ is an encryption mechanism securing stored data. Evidence collection model m in $\mathcal{T}$ specifies the flows of execution, by combining mechanisms in ToC. Evidence ev in $\mathcal{T}$ is expressed in terms of generic test categories and types, and using input/output partitions instead of real data. Life cycle l_is fully defined by the CA in terms of states and transition conditions specified in terms of evidence aggregations.

CM Instance $\mathcal{I}$ is the concrete implementation of a given CM Template $\mathcal{T}$. It includes details on configurations and activities to be executed on the ToC with the final scope of collecting evidence. The definition of a CM Instance $\mathcal{I}$ is a joint effort between the accredited lab and the service provider contributing to the establishment of a full chain of trust (see Sect. 3.7). Before executing the CM Instance $\mathcal{I}$, the accredited lab verifies (i) the signature of the $\mathcal{T}$ from which it is generated or to which is claiming conformance, (ii) the correctness of the instantiation procedure and (iii) the correct representation of the ToC. It is defined as follows.

Definition 3.4.3 (*CM Instance $\mathcal{I}$*) A CM Instance $\mathcal{I}$ is a 5-tuple of the form $\langle \overline{p}, \overline{ToC}, \overline{m}, \overline{ev}, \overline{l} \rangle$. It is the implementation of a given CM Template $\mathcal{T}$, which is signed by a CA. It contains (i) $\overline{ToC}$ as the concrete incarnation of a given ToC, where all the configurations for the real implemented mechanisms execution are specified, (ii) the evidence collection process $\overline{m}$ that can be executed thanks to the configuration in $\overline{ToC}$, (iii) the collected evidence $\overline{ev}$, (iv) the certificate life cycle $\overline{l}$ evaluating the evidence with respect to the certificate status.

We note that the concrete evidence $\overline{ev} \in \mathcal{I}$ is defined in terms of category and types, and in terms of real input and expected output according to the partitions expressed in $ev \in \mathcal{T}$. The evidence $\overline{ev}$ also refers to the real mechanisms $\theta \in \overline{ToC}$, so that executing the concrete collection model $\overline{m}$, they can be re-evaluated if needed (e.g., in case of re-certification).

Example 3.4.2 (*$\mathcal{I}$*) Let us consider CM Template $\mathcal{T}$ in Example 3.4.1. A CM Instance $\mathcal{I}$ for $\mathcal{T}$ has the following elements. Security property $\overline{p} = (Confidentiality, \{\texttt{ctx} = in\text{-}transit/at\text{-}rest\})$. Target of certification $\overline{ToC} = (\{\theta_1, \theta_2\}, <service>)$, where $\theta_1 = (encryption, \{\texttt{algo} = ssh, \texttt{protocol} = TLS1.3, \texttt{level} = message\text{-}in\text{-}transit\})$ implements a secure channel mechanism based on TLS1.3, $\theta_2 = (encryption, \{\texttt{level} = at\ rest, \texttt{algo} = encrypted\ DBMS\})$ implements an encrypted DBMS for artifacts storing. The remaining components should be such that: (i) $\overline{m}$ is consistent with m in $\mathcal{T}$, (ii) $\overline{ev}$ maps ev in $\mathcal{T}$, (iii) $\overline{l} = l$.

The final outcome of a certification process is a certificate. It plays a fundamental role in improving the trustworthiness of cloud services/systems adoption and is an enabler for advanced processes such as certification-aware service selection/discovery and composition. A certificate $\mathcal{C}$ is a 4-tuple $\langle \mathcal{I}, \mathcal{T}, ws, \overline{ev}_r \rangle$, where $\mathcal{I}$ is a CM Instance, $\mathcal{T}$ the CM Template used to instantiate $\mathcal{I}$, ws is the service endpoint where the certificate is shared, and $\overline{ev}_r$ the results of the execution of test cases in $\overline{ev} \in \mathcal{I}$.

We note that according to Fig. 3.3, both the CM Template and the CM Instance are capable of adapting themselves to changes supporting incremental certification (see Sect. 3.5).

3.4.2.2 Certification Process

Figure 3.4 shows our certification process, which is based on the conceptual framework in Fig. 3.3. It is concretely implemented and composed of five main steps.

1. *CM Template Definition*: the CA designs a CM Template $\mathcal{T}$ detailing the methodology for certifying p on a class of *ToC*. We note that this definition might be carried out well before the execution of the other steps of the certification process. It eases the involvement of the CA in dynamic cloud certification since its intervention is decoupled from the process execution.
2. *Service Implementation*: the service provider implements the service *ws* to be certified eventually considering the CM Template as a guide for a certification-oriented design.
3. *CM Instance Definition*: the accredited lab, with the help of the service provider, produces CM Instance $\mathcal{I}$ for the implemented service using the CM Template as a guide. We note that CM Instance can also be defined independently, but have to show conformance to a specific CM Template.
4. *Consistency Check* ($\mathcal{I} \triangleright \mathcal{T}$): the accredited lab verifies the consistency between $\mathcal{I}$ and $\mathcal{T}$. This is crucial to build the certification chain of trust for the cloud. It verifies whether a CM Instance $\mathcal{I}$ is a correct and valid instantiation of a CM Template $\mathcal{T}$. This step is mandatory both when $\mathcal{I}$ is generated independently by the service provider owning the ToC, or if it is generated with the support of the accredited lab.
5. *CM Instance Execution*: the accredited lab executes all certification activities aimed to first certificate issuing (*issuing phase*) and then certificate adaptation (*post-issuing phase*).

We note that the proposed certification process improves trustworthiness via delegation of trust to the accredited lab and supporting consistency check function $\triangleright$, verifying the compliance between what was signed by the CA and executed by the accredited lab. It also provides a dynamic and run-time certificate life cycle management suitable for the cloud.

3.5   Incremental Certification

The initial certificate issuing is often executed in a laboratory environment under the supervision of the CA. In some specific situations, it can be executed in an operation environment by the accredited lab (e.g., in case the property depends on the system operation – configurations and conditions – that cannot be replicated in laboratory). Incremental certification operates after the certificate issuing, to update the certificate at system changes by re-executing part of the certification process. It is clear that incrementality is a crucial property for a cloud certification scheme. More specifically incremental certification aims to maximize certificate validity, while minimizing the risk of unnecessary certificate revocation, reducing as much as possible the amount of re-certification activities. A certificate revocation in fact requires re-certification from scratch, which introduces high cost and time overheads invalidating the benefit introduced by cloud certification schemes.

3.5.1 Certificate Adaptation

CM Instance and CM Template adapt to different types of changes affecting the cloud environment, thus supporting incremental certification.

We consider two adaptation scenarios: (i) *CM Instance adaptation* in case of service, platform, or infrastructure changes, or any changes in the configurations at one or more layers specified in the *ToC*, for instance to cope with elastic scaling or migration; (ii) *CM Template adaptation* in case of new regulations, knowledge, or requirements to assess the validity of a given property (e.g., new issues affecting a mechanism in the *ToC* or new threats). We note that any changes in CM Template also trigger an adaptation process on the corresponding CM Instance. Certificate adaptation provides the ability to re-execute (part of) the process in Fig. 3.4, following changes in the CM Template, the CM Instance, and the service implementation.

3.5.1.1 CM Instance Adaptation
Let us consider an adapted CM Instance $\mathcal{I}' = \langle \overline{p}', \overline{ToC}', \overline{m}', \overline{ev}', \overline{l}' \rangle$ of $\mathcal{I} = \langle \overline{p}, \overline{ToC}, \overline{m}, \overline{ev}, \overline{l} \rangle$. We consider three possible reactions to changes at $\mathcal{I}$ level:
Partial re-evaluation. It applies when the modified $\mathcal{I}'$ is still consistent with the original $\mathcal{T}$. It requires the execution of a partial evidence collection process based on the differences between $\mathcal{I}$ and $\mathcal{I}'$. If possible, it leads to certificate renewal according to the following scenarios.

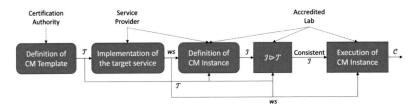

Fig. 3.4 Certification process: execution steps. Rounded boxes represent steps requiring tools or guidelines for service providers, accredited labs, or CA. Squared boxes represent concrete executable tools that have to be implemented

- *Cloud event.* The occurrence of an event affects a mechanism $\theta_i \in \overline{ToC}'$ of $\mathcal{I}$. Test cases involving θ_i are executed again.
- Additional test cases. Additional test cases added in $\mathcal{I}'$, due to a change in $\overline{ToC}'$, $\overline{m}'$, or $\overline{ev}'$, and executed.
- New mechanism. An existing mechanism $\theta_i \in \overline{ToC}$ of $\mathcal{I}$ is replaced by a new mechanism $\theta_j \in \overline{ToC}'$ of $\mathcal{I}'$, having the same type $\theta_j.\hat{\theta} = \theta_i.\hat{\theta}$. Test cases involving θ_i in $\mathcal{I}$ are executed again according to the new $\overline{ToC}'$ and $\overline{m}'$.

If enough correct evidence is collected, the certificate returns to the issued state. On the contrary, partial re-certification or revocation can be triggered. We note that, since the certification authority intervention is not needed, *partial re-evaluation* can be executed at run-time following $\mathcal{I}'$.

Partial re-certification. It applies when the modified $\mathcal{I}'$ is no more consistent with the original $\mathcal{T}$. It requires the execution of a partial evidence collection process that first searches for a new $\mathcal{T}'$ showing consistency with $\mathcal{I}'$, and then follows the differences between $\mathcal{I}$ and $\mathcal{I}'$. The focus of partial re-certification, rather than implementing a complete re-certification from scratch, is to exercise the flows in $\mathcal{I}'$ that either do not exist or differ from the ones in $\mathcal{I}$. New test cases are then executed to collect the evidence needed to award a certificate for a new property $\bar{p}'$ in $\mathcal{I}'$. If the evidence is both correct according to $\mathcal{T}'$ and sufficient according to $\bar{l}$, the certificate is renewed, otherwise the certificate is revoked. As a lightweight alternative, CA defines suitable CM Template $\mathcal{T}'$, such that a weaker property is supported by the service according to the adapted CM Instance $\mathcal{I}'$(i.e., Certificate Downgrade [33]). We also note that a certificate that was downgraded can be then upgraded in case the partial re-certification succeeds.

Full re-certification. It is applied in case changes to $\mathcal{I}$ cannot be managed according to one of the above approaches.

3.5.1.2 CM Template Adaptation

CM Template adaptation focuses on partial updates of the certification methodology. It is driven by the CA that releases a refined CM Template $\mathcal{T}' = \langle p', ToC', m', ev', l' \rangle$ of a given $\mathcal{T} = \langle p, ToC, m, ev, l \rangle$, triggering a CM Instance adaptation for all instances $\mathcal{I}$ referring to $\mathcal{T}$.

CM Template adaptation is a way to support certification-aware resilience to changes. For instance, let Us suppose that a new vulnerability for a given *ToC* is identified, which requires a modification of $\mathcal{T}$. Such modification triggers an adaptation process requiring that all certificates referring to the affected template become immediately *suspended*. The service provider must then adapt the affected services and corresponding instances $\mathcal{I}$ to keep the certificate valid.

3.5.2 Certification Process Adaptation

More recently, incremental certification approaches [44–46] focused on improving how to handle re-certification efficiently. In fact, traditional re-certification techniques failed in representing the system evolution, being based on static models and on static detection of system changes, resulting in an inaccurate planning of re-certification activities. The work in [44] is the first attempt to add dynamicity into the certification process by modeling the system behavior using Machine Learning (ML) models. ML models are used to evaluate the impact of changes and drive certification adaptation. This approach reduces the amount of unnecessary re-certification as follows. It first defines a certification scheme where system behavior modeling tracks system changes over time. It then extends the notion of certification life cycle in Fig. 3.2 by monitoring the system executions with the scope of triggering fine-grained re-certification. More specifically to better track changes, CM Template $\mathcal{T}$ in Definition 3.4.2 is extended to consider the time so that $\mathcal{T}_t$ at time instant t is defined as $\langle p,$ ToC, ev, $\mathcal{T}_t \rangle$, where $\mathcal{T}_{t-1}$ is the reference to the CM Template at time instant $t - 1$. The certificate C_t itself becomes a function of time and is awarded according to $\mathcal{T}_t$. It includes (i) collected evidence $\overline{ev}_t$, including the new evidence collected at time instant t and the subset of evidence $\overline{ev}_{t-1} \in C_{t-1}$ not superseded by evidence in $\overline{ev}_t$; and (ii) the model $\mathcal{B}$ of the system behavior generated by the CA during the initial certification and updated during system evolution.

The adaptation process starts with the initial CM Template $\mathcal{T}_0$ and the corresponding ML-based model $\mathcal{B}$ of the system behavior, as well as the certificate C_0 released by the CA. It then triggers a MAPE (Monitoring, Analysis, Planning, Execute) loop to monitor and detect changes using the system behavior model $\mathcal{B}$ (e.g., detects changes caused by workload variations, code changes, new vulnerabilities). When changes are retrieved, the certificate is suspended. The changes are then analyzed in detail to evaluate their impact on the $\mathcal{T}_{t-1}$ according to the following categories:

- *no impact*: certificate C_{t-1} is still valid (no changes observed) or certification model $\mathcal{T}_{t-1}$ is still valid meaning that changes at time t are minor, that is, the certification model correctly represents the system. In this case, it is only necessary to re-collect some evidence to preserve certificate validity;
- *partial impact*: certification model $\mathcal{T}_{t-1}$ must be revised, since system undergoes some non-negligible changes at time t. For instance, some portions of ToC and p in $\mathcal{T}_{t-1}$ no longer represent the target of certification and property. The certification model needs to be adjusted by the CA and a new release of an updated certificate is requested;
- *full impact*: a new $\mathcal{T}_t$ is requested to correctly represents the system. A set of adaptive actions are executed to generate a $\mathcal{T}_t$ that correctly represents the system at time t.

The above adaptive actions produce a valid $\mathcal{T}_t$ used to collect fresh evidence. In case a positive evaluation is retrieved, a valid Certificate C_t is released including the updated system behavior $\mathcal{B}$, otherwise it is revoked. The key element of the above MAPE process is the model $\mathcal{B}$ of the system behavior, used to detect changes to be later analyzed to evaluate their impact (if any). This approach is capable to overcome the great number of false positive changes detected by the traditional systems [47], having no impact on the system behavior and property to be certified.

3.6 Certification Framework

We describe a certification framework implementing the process in Fig. 3.4, which is suitable for an accredited lab. It provides a set of tools and guidelines for designing CM Templates, generating CM Instances, executing consistency checks, and in general supporting certification management. It is based on two main components: (i) *Certification Model Manager* for CM generation, selection, verification, (ii) *Evidence Collector Manager* to parse the CM Instance $\mathcal{I}$ and drive the evidence collection process by injecting evidence $\overline{ev}$ (i.e., test cases) in $\overline{ToC}$. The proposed framework is generic and can handle different evidence types using type-specific probes to carry out the inspection (e.g., executing test cases).

Certification Model Manager is the main interface for the accredited labs to manage the certification process. It includes a repository of CM Templates and CM Instances, and supports the building blocks in Sects. 3.3 and 3.2, like the hierarchies of properties $\mathcal{H}_P$ and mechanisms $\mathcal{H}_M$. It also includes the *Probe Repository*, where all the probes (e.g., test cases) and the corresponding evidence are stored and kept up to date. It offers internal functionalities such as the consistency check function $f^{\triangleright}$ and the certificate adaptation procedures to support the management of the certification process and certificate life cycle (step $\mathcal{I} \triangleright \mathcal{T}$ in Fig. 3.4).

Evidence Collector Manager implements the evidence collection architecture driven by CM Instance $\mathcal{I}$ [48] (step *CM Instance Execution* in Fig. 3.4). It is composed of a *Probe Manager* (PM), the owner of the collection process, and one or more *Probe Agents* (PAs), responsible for verifying specific flows $\Phi(\overline{m})$ over $\overline{ToC}$. PM (i) provides interfaces to manage the collection process, (ii) configures the evidence collection process, (iii) initializes PAs according to configurations, models, and probes (e.g., test cases) in $\mathcal{I}$, and (iv) manages the certificate life cycle according to $\overline{l}$. PA uses the *Probes Repository* to retrieve the relevant probes in $\overline{ev}$ and execute them on the ToC, returning the results of their execution to PM.

Figure 3.5 shows the data flow between *Certification Model Manager* and *Evidence Collection Manager*. A certification process starts with the *Certification Model Manager* sending a valid CM Instance $\mathcal{I}$ to PM. Upon receiving $\mathcal{I}$, PM parses it and forwards each of its elements to PA(s). The PA(s) retrieves probes (e.g., including the test cases in $\overline{ev}$ of $\mathcal{I}$) and manages the verification activities. The verification activities are executed by one or more probes that access the ToC through a hook. PA(s) sends collected evidence back to PM, when available, which aggregates it and eventually triggers a life cycle transition.

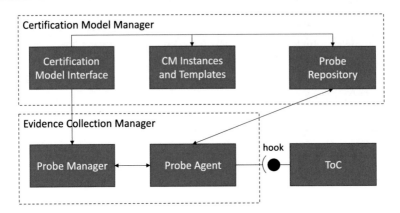

Fig. 3.5 Certification framework architecture

3.7   Certification Chain of Trust

The definition of a proper trust model is fundamental for the usability of a cloud certification process in real scenarios. In traditional software certification, the chain of trust focuses on the role of the certification authorities. In the following, we present a detailed description of the chain of trust for cloud service certification extending the one for software certification.

3.7.1   Chain of Trust

Differently from what happens in classic software certification schemes, in cloud certification the certification authority cannot be assumed to be a single trusted entity taking responsibility for (i.e., digitally signing) the whole certification process. There is a clear need to define a new *chain of trust* whose responsibilities spread across all involved entities and the entire certification life cycle. We therefore design a chain of trust based on multiple signatures. We denote with $A_{en,ws}$ the assertions made by an entity *en* over a system/service *ws*, and with $E_{en,ws}$ the evidence produced by an entity *en* over *ws* and supporting the assertions $A_{en,ws}$. The customer *c*'s trust in an assertion $A_{en,ws}$ made by an entity *en* is denoted $\tau_c(A_{en,ws})$, where τ_c takes discrete values on an ordinal scale (e.g., for a Common Criteria [13] certified product, an assurance level value in 1–7).

Therefore, the minimal chain of trust for cloud certification is based on three different signatures, one for each of the fundamental certification artifacts (CM Template, CM Instance, Certificate) as follows:

- *CM Template signature*: Being the declarative description of the methodology for the certification of a class of *ToC*, CM Template T is signed by a trusted certification authority

CA. $\tau_{AL}(\mathcal{T}_{CA})$ denotes the trust an accredited lab AL has in $\mathcal{T}$ that builds on the trust AL has on CA and its signature.

- *CM Instance signature*: Being the procedural description of the certification activities to be executed, CM Instance $\mathcal{I}$ is signed by an accredited lab AL. The AL is responsible for instantiating the CM Template and, for this reason, it has been delegated by CA. AL receives a signed $\mathcal{T}$ and instantiates it (e.g., filling in all missing elements possibly with the help of the cloud/service providers) to form a CM Instance. The CM Instance signature builds on $\tau_{AL}(\mathcal{T}_{CA})$ and is at the basis of the trust $\tau_c(A_{AL,ws})$ and $\tau_c(E_{AL,ws})$ a client c has in assertions $A_{AL,ws}$ and evidence $E_{AL,ws}$, respectively, provided by AL.
- *Certificate signature*: Being the final document awarded to the service/system, its signature made by the CA binds assertions $A_{AL,ws}$, evidence $E_{AL,ws}$, and the corresponding CM Instance used to carry out certification activities on the *ToC*.

Figure 3.6 shows our minimal chain of trust, where roles are represented with rectangles and artifacts with rounded rectangles. Figure 3.6 also shows certification activities with solid arrows, and trust relations with dashed arrows. This chain of trust models client c's trust in (i) CM Instance $\mathcal{I}_{AL,ws}$, denoted as $\tau_c(\mathcal{I}_{AL,ws})$, used to collect the evidence supporting a set of assertions, (ii) the evidence generated by AL according to CM Instance $\mathcal{I}$, denoted as $\tau_c(E_{AL,ws})$, (iii) assertions made by AL on a service, denoted as $\tau_c(A_{AL,ws})$, where $A_{AL,ws}$ is the set of assertions produced by the accredited lab AL on ws, and (iv) the certificate $\mathcal{C}$ including $A_{AL,ws}$ and $E_{AL,ws}$ and $\mathcal{I}_{AL,ws}$.

$\tau_c(\mathcal{C})$ depends on (i) the reputation of CA signing CM Template $\mathcal{T}$ (i.e., $\tau_{AL}(\mathcal{T}_{CA})$) and the certificate $\mathcal{C}$ itself, (ii) the reputation of AL and the trust in the methodology used by AL

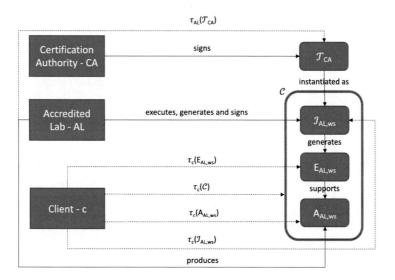

Fig. 3.6 Chain of trust for cloud certification

to generate and sign CM Instance $\mathcal{I}_{AL,ws}$ (i.e., $\tau_c(\mathcal{I}_{AL,ws})$), and specify assertions $A_{AL,ws}$ (i.e., $T\tau_c(A_{AL,ws})$), and (iii) the trust in the methodology used by AL to generate evidence $E_{AL,ws}$ (i.e., $\tau_c(E_{AL,ws})$).

3.7.2   Chain of Trust and Life Cycle Management

The chain of trust is meant to support all phases of the certificate life cycle in Fig. 3.2. Let us start with the certificate issuing phase. It is traditionally based on static evidence and the signature of the CM Instance follows a two-step process: (i) signature of a CM Instance with endpoints that refer to mechanisms deployed in the laboratory, (ii) when the certified service ws is moved in production and certificate C_{ws} is issued (i.e., transition from NI to I in the life cycle), the CM Instance is substituted with a new one signed by AL, with all bindings and endpoints associated with the deployment infrastructure.

The chain of trust also supports following system evolution and cloud events as discussed in Sect. 3.5.1. In this scenario, the collected evidence may become insufficient or contradictory to support one or more assertions, and the corresponding certificate invalid, requiring re-certification. The incremental certification process described in Sect. 3.5 focuses on renewing a certificate by reusing, as much as possible, the certification evidence available from previous certificates [49], reducing the need of new evidence. The trust in this incremental process is given by $\tau_c(E_{AL,ws})$, that is, the trust client c has in the evidence E produced by executing CM Instance $\mathcal{I}_{AL,ws}$, and $\tau_c(\mathcal{I}_{AL,ws})$, that is, the trust client c has in the $\mathcal{I}$. The evolving certificate generated by the incremental process is then handled by the life cycle (e.g., suspension or revocation in the case of insufficient or contradictory evidence). The involvement of the certification authority is marginal according to this chain of trust. It needs to sign the adapted certificate if requested by the accredited lab. The accredited lab delegated by the certification authority is more involved than CA in the process. The CA trusts and verifies lab activities by means of digital signature verification. In case $\mathcal{I}$ is no longer usable for service certification or compliance with the original T is not guaranteed, re-certification from scratch is triggered thus involving the CA.

References

1. P. Stephanow, G. Srivastava, and J. Schütte, "Test-Based Cloud Service Certification of Opportunistic Providers," in *Proc. of IEEE CLOUD 2016*, San Francisco, CA, USA, June, July 2016.
2. I. Kunz and P. Stephanow, "A Process Model to Support Continuous Certification of Cloud Services," in *Proc. of IEEE AINA 2017*, Taipei, Taiwan, March 2017.
3. M. Anisetti, C. A. Ardagna, E. Damiani, and F. Gaudenzi, "A semi-automatic and trustworthy scheme for continuous cloud service certification," *IEEE Transactions on Services Computing*, vol. 13, no. 1, 2020.

4. P. Stephanow and N. Fallenbeck, "Towards Continuous Certification of Infrastructure-as-a-Service Using Low-Level Metrics," in *Proc. of IEEE UIC-ATC-ScalCom*, Beijing, China, August 2015.

5. S. Lins, S. Schneider, J. Szefer, S. Ibraheem, and A. Ali, "Designing Monitoring Systems for Continuous Certification of Cloud Services: Deriving Meta-requirements and Design Guidelines," *Comm. of the Association for Information Systems*, vol. 44, 2019.

6. M. Alhamad, T. Dillon, and E. Chang, "SLA-Based Trust Model for Cloud Computing," in *Proc. of NBIS 2010*, Takayama, Japan, September 2010.

7. Y. Du, X. Wang, L. Ai, and X. Li, "Dynamic Selection of Services under Temporal Constraints in Cloud Computing," in *Proc. of ICEBE 2012*, Hangzhou, China, September 2012.

8. R. Shaikh and M. Sasikumar, "Trust Model for Measuring Security Strength of Cloud Computing Service," *Procedia Computer Science*, vol. 45, 2015.

9. M. Eisa, M. Younas, K. Basu, and H. Zhu, "Trends and Directions in Cloud Service Selection," in *Proc. of IEEE SOSE 2016*, Oxford, UK, March, April 2016.

10. F. Jrad, J. Tao, A. Streit, R. Knapper, and C. Flath, "A utility/based approach for customised cloud service selection," *International Journal of Computational Science and Engineering*, vol. 10, 2015.

11. S. Ding, Z. Wang, D. Wu, and D. L. Olson, "Utilizing customer satisfaction in ranking prediction for personalized cloud service selection," *Decision Support Systems*, vol. 93, 2017.

12. T. Halabi and M. Bellaiche, "A broker-based framework for standardization and management of Cloud Security-SLAs," *Computers & Security*, vol. 75, 2018.

13. D. Herrmann, *Using the Common Criteria for IT security evaluation*. Auerbach Publications, 2002.

14. C. A. Ardagna, R. Asal, E. Damiani, and Q. H. Vu, "From security to assurance in the cloud: A survey," *ACM Computing Surveys*, vol. 48, no. 1, pp. 1–50, 2015.

15. M. Anisetti, C. Ardagna, E. Damiani, and F. Saonara, "A Test-based Security Certification Scheme for Web Services," *ACM Transactions on the Web*, vol. 7, no. 2, May 2013.

16. M. Anisetti, C. Ardagna, and E. Damiani, "Certifying security and privacy properties in the internet of services," in *Trustworthy Internet*, N. B.-M. L. Salgarelli, G. Bianchi, Ed. Springer, 2011.

17. M. Anisetti, C. A. Ardagna, E. Damiani, A. Maña, G. Spanoudakis, L. Pino, and H. Koshutanski, "Security certification for the cloud: The cumulus approach," *Guide to Security Assurance for Cloud Computing*, pp. 111–137, 2015.

18. M. Shanahan, "The event calculus explained," in *Artificial intelligence today: Recent trends and developments*. Springer, 2001, pp. 409–430.

19. G. Spanoudakis, C. Kloukinas, and K. Mahbub, "The serenity runtime monitoring framework," *Security and Dependability for Ambient Intelligence*, pp. 213–237, 2009.

20. A. J. Muñoz-Gallego and J. Lopez, "A Security Pattern for Cloud service certification," in *Proc. of SugarLoaf PLoP 2018*, Valparaiso, Chile, November 2018.

21. S. Gürgens, P. Ochsenschläger, and C. Rudolph, "On a formal framework for security properties," *Computer Standards & Interfaces*, vol. 27, no. 5, pp. 457–466, 2005.

22. A. B. de Oliveira Dantas, F. H. de Carvalho Junior, and L. S. Barbosa, "A component-based framework for certification of components in a cloud of HPC services," *Science of Computer Programming*, vol. 191, 2020.

23. A. J. H. Simons and R. Lefticaru, "A verified and optimized Stream X-Machine testing method, with application to cloud service certification," *Software Testing, Verification and Reliability*, vol. 30, no. 3, 2020.

24. C. Redl, I. Breskovic, I. Brandic, and S. Dustdar, "Automatic SLA Matching and Provider Selection in Grid and Cloud Computing Markets," in *Proc. of ACM/IEEE Grid 2012*, Beijing, China, September 2012.

25. A. Taha, R. Trapero, J. Luna, and N. Suri, "A Framework for Ranking Cloud Security Services," in *Proc. of IEEE SCC 2017*, Honolulu, HI, USA, September 2017.
26. A. Taha, S. Manzoor, and N. Suri, "SLA-Based Service Selection for Multi-Cloud Environments," in *Proc. of IEEE EDGE 2017*, Honolulu, HI, USA, September 2017.
27. K. J. Modi and S. Garg, "A QoS-based approach for cloud-service matchmaking, selection and composition using the Semantic Web," *Journal of Systems and Information Technology*, vol. 21, no. 1, Jan 2019.
28. D. Cotroneo, L. De Simone, and R. Natella, "Dependability Certification Guidelines for NFVIs through Fault Injection," in *Proc. of IEEE ISSREW 2018*, Memphis, TN, USA, October 2018.
29. M. Anisetti, C. A. Ardagna, F. Berto, and E. Damiani, "A security certification scheme for information-centric networks," *IEEE Transactions on Network and Service Management*, vol. 19, no. 3, pp. 2397–2408, 2022.
30. F. Nawab, "WedgeChain: A Trusted Edge-Cloud Store With Asynchronous (Lazy) Trust," in *Proc. of IEEE ICDE 2021*, Chania, Greece, April 2021.
31. R. Calinescu, D. Weyns, S. Gerasimou, M. U. Iftikhar, I. Habli, and T. Kelly, "Engineering Trustworthy Self-Adaptive Software with Dynamic Assurance Cases," *IEEE Transactions on Software Engineering*, vol. 44, no. 11, 2018.
32. S. Jahan, I. Riley, C. Walter, R. F. Gamble, M. Pasco, P. K. McKinley, and B. H. Cheng, "MAPE-K/MAPE-SAC: An interaction framework for adaptive systems with security assurance cases," *Future Generation Computer Systems*, vol. 109, 2020.
33. M. Krotsiani, G. Spanoudakis, and K. Mahbub, "Incremental certification of cloud services," in *Proc. of SECURWARE*, 2013.
34. M. Krotsiani and G. Spanoudakis, "Continuous certification of non-repudiation in cloud storage services," in *Proc. of TSCloud 2014*, 2014.
35. M. Anisetti, C. A. Ardagna, E. Damiani, F. Frati, H. A. Müller, and A. Pahlevan, "Web service assurance: The notion and the issues," *Future Internet*, vol. 4, no. 1, pp. 92–109, 2012.
36. R. Focardi, R. Gorrieri, and F. Martinelli, "Classification of security properties (Part II: Network security)," in *Foundations of Security Analysis and Design II - Tutorial Lectures*, R. Focardi and R. Gorrieri, Eds. Springer Berlin / Heidelberg, 2004.
37. C. Irvine and T. Levin, "Toward a taxonomy and costing method for security services," in *Proc. of ACSAC 1999*, Phoenix, AZ, USA, December 1999.
38. L. Chung, B. Nixon, E. Yu, and J. Mylopoulos, *Non-Functional Requirements in Software Engineering, vol. 5*. Springer, Heidelberg, 2000.
39. L. Chung and B. Nixon, "Dealing with non-functional requirements: Three experimental studies of a process-oriented approach," in *Proc. of ICSE 1995*, Seattle, WA, USA, April 1995.
40. L. Chung and J. Leite, "On non-functional requirements in software engineering," in *Conceptual Modeling: Foundations and Applications*, A. T. Borgida, V. K. Chaudhri, P. Giorgini, and E. S. Yu, Eds. Berlin, Heidelberg: Springer-Verlag, 2009, pp. 363–379.
41. *Health Insurance Portability and Accountability Act (HIPAA)*, U.S. Department of Health & Human Services, November 2015, http://www.hhs.gov/ocr/privacy/hipaa/understanding/.
42. *ISO/IEC 27001 - Information security management*, ISO/IEC, November 2015, http://www.iso.org/iso/home/standards/management-standards/iso27001.htm.
43. *CSA Security, Trust & Assurance Registry (STAR)*, Cloud Security Alliance (CSA), https://cloudsecurityalliance.org/star/, Accessed in Date February 2015.
44. M. Anisetti, C. Ardagna, and N. Bena, "Continuous certification of non-functional properties across system changes," in *Proc. ICSOC 2023*. Rome, Italy: Springer Nature Switzerland, November-December 2023, pp. 3–18.
45. R. Faqeh, C. Fetzer, H. Hermanns, J. Hoffmann, M. Klauck, M. A. Köhl, M. Steinmetz, and C. Weidenbach, "Towards Dynamic Dependable Systems Through Evidence-Based Continuous Certification," in *Proc. of ISoLA 2020*, Rhodes, Greece, October 2020.

46. C. Baron and V. Louis, "Towards a continuous certification of safety-critical avionics software," *Computers in Industry*, vol. 125, 2021.
47. D. S. Herrmann, *Using the Common Criteria for IT security evaluation*. CRC Press, 2002.
48. M. Anisetti, C. A. Ardagna, F. Gaudenzi, and E. Damiani, "A certification framework for cloud-based services," in *Proc. of SAC 2016*, 2016, pp. 440–447.
49. M. Anisetti, C. A. Ardagna, and E. Damiani, "A low-cost security certification scheme for evolving services," in *Proc. of ICWS 2012*. IEEE, 2012, pp. 122–129.

Certification of Modern Distributed Systems

4

Modern distributed systems are increasingly based on composite nano-services deployed in complex dynamic environments (e.g., edge-cloud continuum). In this scenario new requirements emerge aiming to (i) increase the quality of the certificate considering multiple factors affecting the final non-functional property to be certified (Sect. 4.1), (ii) handle certification of micro- and nano-services compositions (Sect. 4.2), (iii) consider peculiarities of the deployment environment while releasing certificates (Sect. 4.3), and (iv) include the development process as a specific target of the certification (Sect. 4.4).

4.1   Multi-factor Certification

The advantages brought by traditional certification schemes conflict with a strong assumption that limits their quality. Traditional certification assumes that certificates are awarded to services based on their software artifacts only, limiting the variety of evidence that can be collected and thus reducing the quality of the certificates.

The certification of modern distributed systems should consider more factors impacting the certificate quality (e.g., information about the programming language used and the type of the development process adopted).

This idea was initially considered by some certification schemes built on hybrid evidence collection models, using a combination of testing, monitoring, and formal proofs [1]. These schemes, although considering evidence of different types thus improving evidence variety, were not capable to address different factors beyond the simple software artifacts. Modern certification schemes extend the verification of a given *ToC* by integrating different aspects influencing the certification. For instance, the development process has a substantial impact on the resulting software and, in turn, on the non/functional properties it holds [2, 3].

© The Author(s), under exclusive license to Springer Nature Switzerland AG 2025   41
M. Anisetti et al., *A Journey into Security Certification*, Synthesis Lectures on Information Security, Privacy, and Trust, https://doi.org/10.1007/978-3-031-59724-4_4

4.1.1 Certification Model

In multi-factor certification schemes, the CM Template ($\mathcal{T}^{fac}$) considers different aspects (factors) influencing the evaluation of the service behavior independently. For example, the work in [4] considers three factors (aka dimensions):

- *Software Artifacts*: describing the software artifacts of the target (i.e., traditional certification based on software evidence);
- *Development process*: describing the development process used to implement the target, such as the programming language used, and the type of development process (e.g., Waterfall, DevOps)
- *Verification process*: describing additional details of the verification process used for certification, such as the fact that the verification is continuous, based on trusted entities or self generated.

The multi-factor CM Template $\mathcal{T}^{fac}$ is defined as follows.

Definition 4.1.1 (*Multi-factor CM Template $\mathcal{T}^{fac}$*) The multi-factor CM Template $\mathcal{T}^{fac}$ is a 6-tuple of the form $\langle p, ToC, m,\ ev,\ l,\ \mathcal{F}\rangle$ that extends the traditional CM Template in Definition 3.4.2 to cope with different factors as follows: (i) the set A_p of attributes in property p is reorganized in subset of attributes, one for each factor, (ii) *ToC* is reorganized in subset of mechanisms, one for each factor, (iii) evidence collection model m is extended to contain flows $\Phi(m)$ targeting the factor-specific mechanisms in *ToC*, (iv) evidence ev reflects the changes in m, including evidence of different types (e.g., part of configuration files), (v) life cycle l expresses generic transitions among certificate states, (vi) function $\mathcal{F}$ defines how to combine evidence of the different factors (e.g., conjunction, disjunction) for certificate awarding.

Table 4.1 shows an example of property p in $\mathcal{T}^{fac}$ for five different implementations s_1–s_5 of the *Artifacts Storage* service in Sect. 3.2. Each property specifies three factors: software artifacts, development process, and verification process.

Considering factor software artifacts only, as in traditional certification, CM Templates of services s_1 and s_4 are equal, being capable of operating with three replicas spread in three zones, with a HA protocol of managed type. CM Templates of s_1 and s_4 specify stronger property reliability with respect to the ones of services s_2, s_3, and s_5. However, considering additional factors, that is, development and verification processes, CM templates of s_1 and s_4 are largely different, with stronger certification requirements posed on s_4.

We note that, although CM Templates in Table 4.1 already suggest a partial ordering among them, the concrete instantiation of such templates could refine such ordering according to the instantiated attributes. In fact, as in traditional certification, the multi-factor CM

Table 4.1 Five different implementations s_1–s_5 of the *Artifacts Storage* service of our scenario in Sect. 3.2. Attributes of property *reliability* is specified according to three factors (i.e., traditional software artifacts, the development process and the verification process)

Service	Software artifacts			Development process		Verification process	
	Repl.	Zones	HA prot.	Prog. lang.	Dev. proc.	Trust. contr.	When
s_1	3	3	**Managed**	Object-oriented	Waterfall	No	After
s_2	3	2	**No**	Object-oriented	Waterfall	No	During
s_3	1	1	**Custom**	Imperative	Prototype	Yes	After
s_4	3	3	**Managed**	Functional and OO	DevSecOps	Yes	During
s_5	2	3	**Managed**	Imperative	DevOps	Yes	After

Template $\mathcal{T}^{fac}$ has to be instantiated into a multi-factor CM Instance $\mathcal{I}^{fac}$, which is ready to be executed, as follows.

Definition 4.1.2 (*Multi-factor CM Instance $\mathcal{I}^{fac}$*) The multi-factor CM Instance $\mathcal{I}^{fac}$ is a 6-tuple of the form $\langle \overline{p}, \overline{ToC}, \overline{m}, \overline{ev}, \overline{l}, \overline{\mathcal{F}} \rangle$ representing a concrete implementation of a given CM Template $\mathcal{T}^{fac}$. More specifically (i) $\overline{p}$ includes concrete attributes to be certified, (ii) $\overline{ToC}$ refers to the real mechanisms to be certified for every factor, (iii) evidence collection model $\overline{m}$ refers to factor-specific probes to be executed on $\overline{ToC}$ to collect evidence, (iv) $\overline{ev}$ is the concrete evidence, (v) $\overline{\mathcal{F}}$ is the concrete factor aggregation function expressing how evidence evaluations in each factor are aggregated for certificate awarding, and (vi) life cycle $\overline{l}$ expresses state transitions in terms of concrete conditions that are computed based on $\overline{\mathcal{F}}(\overline{ev})$.

We note that a simple yet effective evaluation function $\overline{\mathcal{F}}$ can be based on the conjunction of all the single-factor evaluations carried out independently [4].

Example 4.1.1 (*Concrete multi-factor property $\overline{p}$*) Let us consider Artifacts Storage service s_4 in Table 4.1 where the property p in the multi-factor CM Template is expressed as $p = (\hat{p}_{rel}, (A_p^{art}, A_p^{dev}, A_p^{eval}))$. We note that the attributes $(A_p^{art}, A_p^{dev}, A_p^{eval}$ are subject to total order relations $\preceq_{a_k}$ between contextual attributes $a_k \in A_p^j$ of a given factor j. This total ordering can be further refined in the corresponding concrete property $\overline{p} \in \mathcal{I}^{fac}$. For instance, an instantiation of the attribute Prog. Lang. (pl) of factor development process (A_p^{dev}) is subject to a total order where [Rust $\preceq_{a_{pl}}$ Java $\preceq_{a_{pl}}$ Python]. According to templates in Table 4.1, s_4 was developed with a more robust programming language (*Prog. Lang. =*

Functional and OO), a more advanced development process (*Dev. Proc.* = *DevSecOps*), and verified using a continuous process (*when.* = *During*).

We note that each factor-specific evaluation in $\overline{m} \in \mathcal{I}^{fac}$ depends on the factor peculiarities, as well as on the target mechanisms. For instance, the evaluation of the development process is carried out by specific probes that inspect the metadata description attached to the service to be certified, where the attributes of the development process are detailed (e.g., the programming language used, and the type of development process). The multi-factor scheme provides benefits for all the parties involved in the certification process as follows. The service provider retrieves certificates better describing the behavior of its services [5–7]. The end user selects services according to more accurate certificates, supporting fully/informed and safe decisions. The CA improves the quality and, in turn, the trustworthiness of its certification scheme, providing higher accuracy with a marginal increase in overhead.

Example 4.1.2 (*Concrete target* $\overline{ToC}$ *and evidence collection model* $\overline{m}$) Let us consider the CM Template of service Artifacts Storage s_4 in Table 4.1. The target $\overline{ToC}$ can be defined as $\overline{ToC} = \{\theta_{art}, \theta_{dev}, \theta_{eval}\}$, where for instance $\theta_{art}.\ A_m = \{Replica\ Manager = \text{Kubernetes}\}$, $\theta_{dev}.\ A_m = \{Pipeline = \text{File Content}, Source\ Code = \text{Rust}, Code\ Review\ Document = \text{File Content}\}$. $\overline{ToC}$ expresses that s_4 is deployed in Kubernetes, is written in Rust, and has a code review document. The evidence collection model $\overline{m}$ contains the description of the execution flows to collect evidence on $\overline{ToC}$ to verify property $\overline{p}$. For instance, evidence collection model for factor artifacts $\overline{m}_{art} = \{(Replica\ Manager = \text{Kubernetes}, \text{Get/Orchestrator}), (Replica\ Manager = \text{Kubernetes}, \text{Check/Replicas}), (Replica\ Manager = \text{Kubernetes}, \text{Check/Zones})\}$, where Get/Orchestrator, Check/Replicas, Check/Zones are different evidence collection flows (i.e., probes) that first retrieve the orchestrator, then verify the number of replicas and zones.

We note that although Example 4.1.2 considers attributes that indirectly refer to the deployment environment (e.g., the Kubernetes cluster), no specific evaluation flows are listed in the evidence collection model $\overline{m}_{art}$ apart form the ones needed to support replicas based on configurations checks. More advanced checks, specifically tailored to the deployment environment, are described in Sect. 4.3.

4.2   Certification of Service Composition

Service composition is bringing the dynamicity of cloud services to an unprecedented level. A service can be dynamically offered to a given client by combining already existing services into a newly composed one. In this context, a certification scheme must consider the whole composition life cycle, including all phases from service selection to the non-functional evaluation of the composite service.

Most of the assurance solutions addressing the composition of services in literature focused on non-functional aware composition generation [8–12] or QoS-aware composition monitoring and building [13–16]. These solutions are not fully suitable in the context of certification due to the lack of evidence and structured models, which must be part of a trustworthy certification methodology. Most of the composition certification approaches presented in literature require intensive involvement of CA and accredited lab [17]. Such involvement collides with the extreme dynamicity of the compositions, which often change at run-time to address specific requirements (e.g., performance). A new emerging trend is the adoption of lightweight composition certification schemes, as the way to support dynamic composition by run-time component substitution [18, 19]. A certification process for composite services requires four fundamental building blocks: (i) composition modeling, driving the functional and non-functional orchestration and handling the dynamic evolution of the composition, (ii) composition annotation, associating non-functional requirements to each component services, (iii) component selection, supporting the selection of suitable components at run-time (i.e., to replace migrated services), and (iv) certification of composite services, awarding certificates for dynamically-generated service compositions.

4.2.1   Composition Modeling

Two types of service compositions exist: (i) orchestrations, where a central entity (the orchestrator), act as a middleware between the component services mediating their communications, (ii) choreography, where services directly exchange messages in the framework of the composition. In the following, we consider orchestration-based compositions. Among the possible languages modeling a composition of services (e.g., OWL-S [20]), BPEL is largely used to support orchestration-based compositions. BPEL is an XML-based language that permits to define a business process by using a collection of services. More specifically it permits to (i) specify the order in which service operations are invoked, (ii) the data to be exchanged at each step of the composition, and (iii) the conditions under which a service is selected and integrated within the business process [21]. BPEL defines executable processes that consist of a set of activities (e.g., *invoke*, *receive*, and *reply*) combined using different structures to form a coherent system. BPEL engine, in most of the cases, supports run-time service composition, where component services, which are listed in a registry, are dynamically selected and composed on the basis of functional requirements. We model the BPEL service composition as a graph as follows.

Definition 4.2.1 (*Composition Graph*) A Composition graph $G(V, E)$ is a direct acyclic graph with a root $v_r \in V$. G has (i) a vertex $v_i \in V_I \subseteq V$ for each service operation invocation, (ii) two additional vertices $v_c, v_m \in V_\otimes \subset V$ for each alternative ($\otimes$) structure modeling the alternative execution (*choice*) of operations and the retrieval (*merge*) of the results, respectively, and (iii) two additional vertices $v_f, v_j \in V_\oplus \subset V$ for each parallel ($\oplus$) struc-

Table 4.2 Services of our scenario in Sect. 3.2

Service	Operation	Description
Compliance Control Repository (CCR)	*Control* `SelectControl`(*ID*)	Permits to select suitable controls to be executed
Evaluation Manager (EM)	*[res, art]* `ExecuteControl`(*Control, target*)	Permits to execute a given control on a given target *target*
	Success `PublishResults`(*res*)	Permits to publish results to the Compliance Manager to trigger compliance evaluation
Assurance Storage (AS)	*Success* `SaveArtifacts`(*data*)	Stores compliance artifacts

ture modeling the contemporary execution (*fork*) of operations and the integration (*join*) of their results, respectively.

We note that $\{v_r\} \cup V_I \cup V_\otimes \cup V_\oplus = V$, and v_c, v_m, v_f, and v_j model branching for alternative/parallel structures. We also note that root vertex v_r represents the BPEL orchestrator, that is, the set of operations exposed in the interface by the process owner. For simplicity, we consider the orchestrator just as a means to trigger services' invocation and therefore out of the scope of the certification (i.e., no internal functionalities to be certified). Other solutions recognizes the importance of the orchestrator (e.g., to keep the integrity of the functional flows of invocation) and involves it in the certification process as first class citizen [22].

Example 4.2.1 (*BPEL Composition*) Let us consider our scenario in Sect. 3.2 about a component-based system evaluating compliance. Let us consider a service composition realizing it, where each component is implemented as a service. Table 4.2 shows services and operations involved. Let also assume to have multiple alternative services implementing the Compliance Control Repository (CCR_1 ... CCR_3) and, for simplicity, one service implementing the Evaluation Manager (EM) and one the Assurance Storage (AS). The Compliance Manager plays the role of BPEL orchestrator. Figure 4.1 shows the BPEL model of this scenario based on a sequence of orchestrated operation calls between services.

4.2.2 Composition Annotation

Annotations have been proposed in the past to support selection of services to be integrated within a composition [9, 10, 12]. Each BPEL invocation can be annotated detailing non-functional property to be satisfied by the component services, and in turn detailing the

Fig. 4.1 BPEL graph of Example 4.2.1 relatedo to our scenario in Sect. 3.2

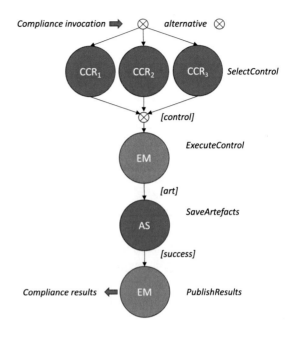

propriety for which it has to be certified for. Annotation can be seen as a labeling function λ : $V_I \rightarrow L_S$ that associates a set of non-functional requirements $\mathcal{R} \in L_S$ with each invocation in $v \in V_I$. We formally define an annotated BPEL graph as follows.

Definition 4.2.2 (*Annotated Composition Graph*) An annotated Composition graph $G^{\lambda}(V, E, \lambda)$ of a given a Composition graph $G(V, E)$, is a direct acyclic graph with a labeling function λ that assigns a label $\lambda(v)$ to each service operation invocation represented as vertex $v \in V_I$. $\lambda(v)$ corresponds to the non-functional property to be satisfied by the component service represented by v.

We note that security annotation $\lambda(v)$ can be expressed using metadata format (e.g., XML), thus modeling requirements in a rich and structured form. For instance, in case the selection have to be done among certified services, it is possible to specify details on the requested property p and evidence ev. Figure 4.2 shows an annotation expressed as XML fragment requesting a service certified for property *Confidentiality* of data *in transit* using SSL/TLS with a key length equal to $2048\ bit$. It also requires an evidence composed of at least 200 test cases of category *Penetration*. The composition graph in Definition 4.2.2 is annotated with the non-functional property p that a service owner wants to certify on its composite service. It is at the basis of *Composition Template* $\mathcal{T}^{com}$.

Definition 4.2.3 (*Composition CM Template* $\mathcal{T}^{com}$) A Composition CM Template $\mathcal{T}^{com}$ is 5-tuple of the form $\langle p, G^{\lambda}(V, E, \lambda), m, ev, l, \rangle$ where (i) p is the property to be certified for

Fig. 4.2 An example of non-functional annotation for a given invocation in the BPEL

```
<annotation id="01" ref="Invocation">
  <property>
    <abs>Confidentiality</abs>
    <algo>SSL/TLS</algo>
    <key>2048 bit</key>
    <ctx>in transit</ctx>
  </property>
  <evidence>
    <category>Penetration</functionality>
    <attrs>
      <attr id="cardinality">200</attr>
    </attrs>
  </evidence>
</annotation>
```

the entire composition, (ii) the annotated graph $G^\lambda(V, E, \lambda)$ is the compositional target of certification mapping p into specific requirements for each component service, (iii) m^f and ev are the same as in traditional CM Template in Definition 3.4.2, but contain component-specific and composition-specific evidence collection models and evidence, respectively, (iv) life cycle l is a traditional life cycle expressed in terms of evidence ev.

Similarly to CM Template the process of defining the Composition CM Template involves (i) the CA to carry out annotation on G^λ and (ii) an accredited lab to verify its validity.

Example 4.2.2 (*Annotated BPEL*) Let us consider BPEL in Example 4.2.1 and a target property *p Confidentiality in transit*. An Annotated BPEL can be defined considering services CCR_i, EM, and AS annotated to support non-functional property *Confidentiality* of data *in transit* using BPEL annotations as defined in Fig. 4.2.

4.2.3 Candidate Service Selection

Our goal is to generate an executable service composition proving a specific non-functional property. To this aim, we create a Compositional Instance $\mathcal{I}^{com}$ that instantiates the Composition Template $\mathcal{T}^{com}$ with component services. Component services are selected and connected according to the template and the target non-functional property. Many solutions exists in literature to select services based on their non-functional properties (e.g., based on SLA annotations [23–25] or certificates [26]).

The service selection process can be considered as a function that takes as input the annotated graph $G^\lambda(V, E, \lambda)$ of a given Composition CM Template $\mathcal{T}^{com}$ and different sets of candidate services, each one satisfying the functional properties of one service operation invocation, and returns as output an executable composition graph $G'(V', E)$ where every invocation $v \in V_I$ contains a service instance v', and every branching $v \in V_\otimes \cup V_\oplus$ is maintained as it is. The executable composition graph $G'(V', E)$ is obtained by traversing the graph G^λ with a *depth-first* search. The algorithm starts from the root vertex v'_r of G' containing the instance of the orchestrator. Then for each vertex $v \in V_\otimes \cup V_\oplus$, a corresponding vertex $v' \in V'_\otimes \cup V'_\oplus$ is generated where each vertex $v' \in V'_I$ is an instance of a real service ws_i, such that the ws_i functionally compatible with the functional requirements in $G^\lambda(V, E, \lambda)$ and satisfies the annotation $\lambda(v)$.

More in details to select the candidate service for v' a two-step candidate selection approach is applied as follows.

1. *Match*: it considers a set S of candidate services s_i and $\lambda(v)$ as the non-functional requirements for the invocation at vertex v in the Composition Template. Assuming that functional matching is successful, non-functional requirements are considered by evaluating whether s_i satisfies $\lambda(v)$ (i.e., s_i owns a certificate C_i that satisfies $\lambda(v)$). The matching algorithm returns a set $S' \subseteq S$ of compatible services, which represent the possible candidates for selection.
2. *Rank*: compatible services $s_i \in S'$ are ranked in a partial or full order on the basis of their non-functional characteristics (e.g., on the basis of their certificates C_i) . The best ranked service is then selected and integrated in $v' \in V'_I$.

We note that steps matching and ranking depend on how the non-functional property of a given component service is expressed both in the annotation and as metadata associated with the component services.

In case of certified component services and multi-factor schemes, matching may need to check all the factors detailed in the certificates (see Sect. 4.1) and the ranking may need to rank based on these factors [4].

The executable composition graph $G'(V', E)$ is used to build the Composition CM Instance $\mathcal{I}^{com}$ as follows.

Definition 4.2.4 (*Composition CM Instance $\mathcal{I}^{com}$*) A Composition CM Instance $\mathcal{I}^{com}$ is a 5-tuple of the form $\langle \overline{p}, G'(V', E), \overline{m}, \overline{ev}, \overline{l} \rangle$ where (i) $\overline{p}$ is the concrete property, (ii) $G'(V', E)$ is the executable composition graph constituting the concrete $\overline{ToC}$, (iii) $\overline{m}$ represents the executable evidence collection activities to be carried out on the composed service whose results (evidence $\overline{ev}$) are evaluated following certificate life cycle $\overline{l}$.

Example 4.2.3 (*Composition CM Instance*) Let us consider the annotated BPEL in Example 4.2.2 and the corresponding Composition CM Template $\mathcal{T}^{com}_s$. For simplicity, let us also assume that three candidate services for AS ($ws_1, \ldots, ws_3$) are available out of the five described in Table 4.1, while for the remaining services we have one candidate service: ws_4 for CCR_1, ws_5 for CCR_2, ws_6 for CCR_3 and ws_7 for EM. Let us assume that each service $ws_5, \ldots, ws_7$ owns a certificate for property *Confidentiality* of data *in transit* showing at least 200 test cases of category *penetration* (see Fig. 4.2), while the certificate for ws_5 also owns a certificate for property *Confidentiality* of data *in transit and at rest* with 400 test cases of category *penetration*. Therefore $ws_4, \ldots, ws_7$ are directly selected by our candidate service selection at phase match, since they are fully compliant with the annotation on the Composition CM Template $\mathcal{T}_s^{com}$ and there are no other candidate services to be ranked. Let us also assume that service ws_1 is certified for *Confidentiality* of data *at rest* without specifying test cases, while service ws_2 and service ws_3 are certified for *Confidentiality* of data *in transit and at rest*, but only service ws_2 owns a certificate with more than 500 penetration tests. Considering our service selection process, the matching phase selects ws_2 and ws_3, while phase rank ranks ws_2 at the top. The Composition CM Instance $\mathcal{I}^{com}$ to be considered for certification is therefore the one composing services $ws_2, ws_4, ws_5, ws_6, ws_7$, and their corresponding certificates.

4.2.4 Composition Certification

Given a composition CM Instance $\mathcal{I}^{com}$, the service provider can (i) decide to certify the entire composition, including each of its components, according to the Composition Template, or (ii) re-use the certificates available for the component services to generate the certificate of the composition. More specifically certification process can proceed as follows.

- *Composition Full-Certification*: The Composition Template and Instance are used to set up a complete certification process from scratch as in Sect. 3.4. It produces a traditional monolithic certificate with evidence generated by the execution of the given $\mathcal{I}^{com}$. This process does not consider existing certificates associated with component services. Specifically, $\overline{m} \in \mathcal{I}^{com}$ models the evidence collection flows targeting each component services. It is used as in traditional certification to carry out testing and retrieve evidence. Collected evidence $\overline{ev}$ is the result of the execution of the different flows in $\overline{m}$ on all software components part of the composition.
- *Composition Re-certification*: Each component services has a stand-alone certificate used in service selection (see Sect. 4.2.3). The Composition Template is considered as the Template for the entire composition. The CM Instance of each component service can be reused or re-executed for composition certification depending on the property to be certified and on the composition structure. Composition re-certification process uses $\overline{m} \in \mathcal{I}^{com}$ and $\overline{ev} \in \mathcal{I}^{com}$ as well as the corresponding $\overline{m} \in \mathcal{I}$ and $\overline{ev} \in \mathcal{I}$ of each component service that is requested to be re-certified. This process generates a certificate where evidence is not generated from scratch as for the Full-Certification, but are selected from the ones available at component level. Such evidence is then just re-verified at each component level if needed (e.g., if involved in specific mandatory flows $\overline{m} \in \mathcal{I}^{com}$ for a given property to be certified).
- *Virtual Composition Certification*: The core idea is to empower CA to virtually certify a composite service starting from the certificates of each component services, without

executing or re-executing any evidence collection activities. This certificate is "virtual", since it does not involve any real evaluation or re-evaluation activities [18]; it is generated by combining the existing certificates of each component according $G'(V', E)$ in $\mathcal{I}^{com}$. It iteratively composes pairs of (virtual) certificates $\mathcal{C}_i$ and $\mathcal{C}_j$ to generate a virtual certificate $\mathcal{C}^*_{G'}$ for the composition. The virtual certification process starts considering services composed in a sequence. All pairs of services in a sequence are selected and their certificate $\mathcal{C}_i$ and $\mathcal{C}_j$ composed into a virtual one $\mathcal{C}^*_{ij}$. Their respective vertices $v_i, v_j \in V'$ are substituted with a single vertex v_{ij} annotated with certificate $\mathcal{C}^*_{ij}$. When no service sequences are left, a pair of services in either an alternative ($\otimes$) or a parallel ($\oplus$) are composed, a virtual certificate $\mathcal{C}^*_{ij}$ generated, and their vertices $v_i, v_j \in V'$ substituted with a single vertex v_{ij} annotated with certificate $\mathcal{C}^*_{ij}$. This process is repeated until G' is reduced to a graph having a single vertex v with virtual certificate $\mathcal{C}^*_{G'}$. Given two certificates $\mathcal{C}_i$ and $\mathcal{C}_j$, a virtual certificate $\mathcal{C}^*_{ij}$ is generated composing the properties, the models and the evidence of $\mathcal{C}_i$ and $\mathcal{C}_j$ as described in [18]. The trust in a virtual certificate is mainly based on the trust a customer has in (i) the automatic process generating the virtual certificate, which is implemented by the accredited lab delegated by the CA and (ii) the certificates held by the individual component services.

Example 4.2.4 (*Virtual Certification*) Let us consider $\mathcal{I}^{com}$s in Example 4.2.3, which is composed of services $ws_2, ws_4, ws_5, ws_6, ws_7$ as presented in Fig. 4.1. In the framework of virtual certification, given the fact that services $ws_4, \ldots, ws_6$ are combined in an alternative $\otimes$, the virtual certificate of alternative services $\mathcal{C}^*_\otimes$ is equivalent to the certificate with the minimum strength, that is, $\mathcal{C}_{ws_5}$ (or $\mathcal{C}_{ws_6}$). Considering virtual certificate $\mathcal{C}^*_\otimes$ and certificates $\mathcal{C}_{ws_2}$ and $\mathcal{C}_{ws_7}$ composed in a sequence, the composition can be awarded with the virtual certificate $\mathcal{C}^*_s \langle \mathcal{I}^{com}, \mathcal{T}^{com}, ws_s, \overline{ev}_r \rangle$ for property (*Confidentiality*, {ctx = *in-transit*}), where $\mathcal{I}^{com}$ is based on the $\mathcal{I}$ of each component.

Let us assume that the certificate $\mathcal{C}_{ws_5}$ in Example 4.2.4 is suspended for property confidentiality in transit and still valid for confidentiality at rest. The entire virtual certification is suspended even if property confidentiality at rest is still valid for $\mathcal{C}_{ws_5}$, since such property is not requested for the certification of the composition.

4.3   Certification of the Deployment Environment

The deployment environment has a substantial impact on the properties of the services executed on it. A clear impact is on the performance boost a deployment platform can guarantee to a service. There is an increasing interest in literature in understanding the impact of different deployment solutions on the non-functional posture of services. Most of the proposed solutions focused on performance-related QoS fulfillment [27, 28], considering simple Function-as-a-Service (FaaS) deployment platforms [29–31] for stateless services. Just few solutions address the security/privacy impact of different deployment solutions [32, 33] considering service workflows. The core idea is that, the deployment infrastructure can either offer security features to strength the deployed services or introduce security weaknesses impacting the service security. A pioneering work was carried out in [34], where properties or web service containers are considered in the framework of web service certification. However, the number of properties supported by this containers certification is very limited. Multi-factor certification schemes, like the one in Sect. 4.1, already partially tackle the deployment environment. For instance in Example 4.1.2, configurations of the Kubernetes deployment environment are involved in factor artifacts for the certification of property reliability. However, since the target of certification is at service layer, most of the infrastructure-specific peculiarities are missing. For instance, specific weaknesses possibly affecting the working of the infrastructure are neglected.

This section is focused on the certification of a deployment environment as the sole target of the certification process, independently from the hosted services. Two approaches are possible: (i) adoption of traditional service certification where the infrastructure is the target of certification, (ii) use an ad hoc approach for the certification of a container [34], focusing on the properties that can be inherited by the services executed within it. While the first approach as a low level of applicability, since the infrastructure is in most of the cases out of the control of the service owner, the container certification is more feasible, but still somehow out of the control of the service owner.

More recently, different approaches have been proposed, having the primarily objective of empowering the deployment environment with hooks exposing certification metrics that make the environment certification-ready [35, 36]. These approaches are even more interesting with the advent of edge solutions that involve telco operators in service deployment [32]. In this scenario, it is not feasible to deploy certification probes on telco nodes or to control them; therefore, available infrastructure-layer metrics are extended and added to the telco standards limiting the performance impact of certification.

4.3.1 Certification Modeling

A certification model targeting a deployment environment is grounded on a different perspective than a certification model for services. The certification of a deployment environment, in fact, would increase its credibility and its attractiveness from a business perspective, definitely opening to the composition of certified services as described in Sect. 4.2. The certification model therefore assumes a more collaborative behavior by the environment owner in providing means of collecting evidence. On the other side, given the multi-tenancy nature of a given deployment environment, it is less worthy to have a plethora of CPU-demanding probes deployed and executed in such a shared environment. In addition, a modern deployment environment is increasingly based on telco facilities such as MEC edge services, which

provide restricted access to external inspections. Given this scenario, the deployment certification model is grounded on the assumption that the environment provides hooks offering
relevant *Metrics* to be aggregated as *Rules*. The rules are meant to evaluate a specific aspect
of a given deployment environment. Rules are aggregated in terms of *Policies* defined by a
CA, each of them aimed to prove a specific behavior supporting a given *non-functional property*. Given the dynamic nature of the metrics and the absence of control on the deployment
platforms, differently from the other certification processes, the time window of metric measurement is considered to indicate the period of validity, as well as to drive the aggregation
in terms of rules.

The CM template $\mathcal{T}^{dep}$ for deployment environment is structured as follows.

Definition 4.3.1 (*Deployment CM Template* $\mathcal{T}^{dep}$) A $\mathcal{T}^{dep}$ is 5-tuple of the form $\langle p, ToC,$
$t, pol, ev, \rangle$ where (i) property p is the property to be certified, (ii) *ToC* is the target of
certification expressed as the different mechanisms at environmental level supporting the
property to be certified, (iii) t is the time interval for the evaluation, and (iv) policies *pol*
represents the generic policies defined in terms of rules to certify the property p. Collected
evidence *ev* represents the expected outcomes of the rules expressed in terms of metrics.

We note that the life cycle is included in the definition of the policy *pol*. As for the other
certification processesm, $\mathcal{T}^{dep}$ needs to be instantiated to be executed by the accredited lab.
Rules and metrics that are part of the policies *pol* in the $\mathcal{T}^{dep}$ need to be concretely specified
on the real target.

Definition 4.3.2 (*Deployment CM Instance* $\mathcal{I}^{dep}$) A $\mathcal{I}^{dep}$ is 5-tuple of the form $\langle \bar{p}, \overline{ToC},$
$t, \overline{pol}, \overline{ev}, \rangle$, where (i) property $\bar{p}$ is the concrete property to be certified, (ii) $\overline{ToC}$ is the
concrete target of certification linking the concrete executable environmental level mecha-
nisms supporting the property to be certified, (iii) t is the time interval for the evaluation,
(iv) policies $\overline{pol}$ represents the concrete policies as a set of concrete rules expressed in terms
of real metrics and aimed to certify the property $\bar{p}$, and (v) collected evidence $\overline{ev}$ represents
the concrete outcomes of rules expressed in terms of metrics.

Let us consider a given $\mathcal{I}^{dep}$ expressing a non-functional property (*Confidentiality*, {ctx
= *in-transit/between nodes*}) to be certified for a target environment $\overline{ToC}$ made of set of
nodes in a continuum. The accredited lab implements the CA's policies in $\mathcal{T}^{dep}$ as $\overline{pol}$ in $\mathcal{I}^{dep}$
in terms of rules aggregating metrics with the aim of collecting the evidence $\overline{ev}$ supporting
the property. Two rules can be defined as follows $r_i(\mathbf{n}, \mathbf{t}) = m(\mathbf{n}, \mathbf{t}) \geq 2048$ where the metric
m indicates the size of the RSA key used to encrypt all communications between the nodes
in $\mathbf{n}$ in a given time interval $\mathbf{t}$. The second rule $r_j(\mathbf{n}, \mathbf{t})$ is true *iff* $\mathbf{n}$ uses valid certificate for
message encryption in the time interval $\mathbf{t}$. Accredited lab evaluates the two rules included
in the policy against a given set of nodes $\mathbf{n}$ and a time interval $\mathbf{t}$. If successful, it issues a
certificate including the policy, the metrics, and the evidence obtained.

Example 4.3.1 (*Certification of Deployment Environment*) Let us consider Example 4.1.2, where Kubernetes is part of a factor for the certification of property reliability. Let us then consider the Kubernetes cluster as the target *ToC* of a deployment environment certification for the same property (*Reliability*, {ctx = *Replication*). The scope of this certification is to verify that the *ToC* supports property reliability via replication. The corresponding $\mathcal{T}^{dep}$ includes generic policies *pol* with rules requiring that the target Kubernetes cluster shows replication of nodes. It is instantiated in the $\overline{pol}$ of the CM Instance $\mathcal{I}^{dep}$, where rules are implemented and metrics collected. For instance, $\overline{pol}$ includes a rule instance $r_1(Kubernates\ cluster, \mathbf{t})$ that is true *iff* the Kubernetes shows a working cluster in HA in the time interval $\mathbf{t}$ of the evaluation. It means that metrics monitoring the availability of the different cluster nodes are provided and collected by rule r_1 in the time interval $\mathbf{t}$. We note that other rules collecting configurations are part of the given $\overline{pol}$ as well.

We note that differently from Example 4.1.2, the certification activities are specifically focused on the Kubernetes cluster and not on the service deployed on it. We also note that some of the evaluation activities in Example 4.1.2 can be also executed at deployment level as soon as the metrics are made available by the infrastructure owner.

4.4   Certification of the Development Process

Certification of development process aims to extend the concept of *continuous certification* with the scope of covering as much as possible the software life cycle from solution design, its coding and continuous updating while in operation. The certification of the development process goes beyond the verification of the final product deployed in operation, and continuously verifies all artifacts produced by the entire development process [37]. The idea is that the quality of the entire development process has an impact on the final results and therefore its certification is of paramount importance.

Traditional approaches focused on the certification of some aspects of the development process only [38, 39], just few took the specific system under development into account [40], and even fewer addresses modern development processes [37, 41, 42]. Certification of development process needs to target all the software artifacts (including the intermediate ones) and bind collected evidence to all development stages, linking evidence back to certification requirements. This process "shift to left" the non-functional activities that usually has been thought as a "bolt-on" process after the provisioning of the application. More in detail, this notion of continuous certification introduces new requirements on certification schemes that are summarized as follows.

- **Development process modeling**: the development process must be part of the certification model. This modeling is under the responsibility of the CA.

- **Stage-by-stage evaluation**: the certification model must describe both requirements and the corresponding controls used to collect relevant evidence at each development stage.
- **Trustworthy inspection**: the development process must support trustworthy execution of controls.
- **Continuous verification**: hooks in the development process must be provided to support controls in continuously collecting evidence.

Although our notion of continuous certification (and corresponding requirements) is applicable to any development processes, it provides the best advantages in agile processes, where Continuous Integration (CI) and Continuous Delivery (CD) make the development process and the software it releases a unique conceptual entity. In the following, we then focus on certification of services developed following a DevOps process, presenting a certification scheme for DevOps. DevOps in fact provides full control and automation of the development steps, and is a natural candidate for supporting continuous certification.

An overview of the activities done at each DevOps stage is sketched in the following.

- Stage *Plan* designs the application. It also specifies security requirements and metrics to measure performance and quality of service of the application itself.
- Stage *Create* generates the application code, statically verifies it for security reasons, and builds it after the source code is committed.
- Stage *Verify* provides features for functional and non-functional testing of the software.
- Stage *Package* prepares the software artifacts for application deployment, and the corresponding security verification on packaging dependencies.
- Stage *Release* marks the difference between staging and operation pipelines targeting private or public deployments, respectively.
- Stage *Configure* configures the deployment environment, and includes infrastructure-level and container-level security verification.
- Stage *Monitoring* continuously monitors the application to calculate those metrics needed for continuous security verification.

We note that stages Release, Configure, and Monitor consider both staging and operation environments.

4.4.1 Certification Modeling

Similarly to traditional certification schemes, certification of a development process is modeled in a CM Template ($\mathcal{T}^{dev}$) that verifies the adherence of the whole development process to relevant policies and requirements. CM Template $\mathcal{T}^{dev}$ triggers specific verification activities C at each stage of the development process. Targets T of these activities are software components (e.g., non-functional mechanisms θ), configuration files, or planning documentation files.

The Development Certification Model Template $\mathcal{T}^{dev}$ is structured as follows.

Definition 4.4.1 (*Development CM Template $\mathcal{T}^{dev}$*) A $\mathcal{T}^{dev}$ is a 4-tuple of the form $<stage_i, T_i^s, C_i, \Upsilon_i>$, where $stage_i$ refers to the i-th DevOps stage triggering the execution of certification activities in C_i, $T_i^s \subseteq T$ is an artifact of the system under certification (e.g., source code, executable, configuration files), $C_i \rightarrow \{c_{i,1}, \ldots, c_{i,n}\}$ is a list of generic control types suitable for system certification, and Υ_i defines the guards that need to be satisfied for triggering a stage transition.

We note that each C_i specifies a class of controls, which point to a set of concrete controls $\{c_{i,1}, \ldots, c_{i,n}\}$ to be used for system verification. For instance, class *web vulnerability* includes the controls that provide web vulnerability assessment functionalities. We also note that guards Υ_i are defined as a boolean conditions of the form $op(\{c, T^s\}, EO)$, where op is an operator in $\{=, \leq, \geq, <, >\}$, c is a control, T^s is the target of the control, and EO is the expected output of the execution of control c on target T^s. Guards express conditions to be satisfied by the specific stage to proceed with the following one. We note that if a guard is not satisfied, the certification process ends and some correction activities might be requested to address the failure. We also note that, for some target artifacts T^s, traditional controls c are used [43].

Figure 4.3 shows an excerpt of a Development Certification Model Template $\mathcal{T}^{dev}$ for the DevOps Process of service Assurance Storage in Example 4.1.2. It is structured as follows. $stage_1$ refers to stage Plan. C_1 contains a single manual control $c_{1,1}$ ensuring that the requirements document exists and has the format requested to drive the remaining of the process (Υ_1).

$stage_2$ refers to stage Create. C_2 contains two types of controls: (i) $c_{2,1}$ verifies whether the requirements document in stage Plan ($T_{2,1}^s$) has been manually annotated to link requirements to the source code components they regulate; (ii) $c_{2,2}$ statically verifies the source code in the code repository ($T_{2,2}^s$) to identify security issues. In case both $c_{2,1}$ and $c_{2,2}$ succeed (Υ_2), stage Verify is executed.

$stage_3$ refers to stage Verify. C_3 contains a single control $c_{3,1}$. It refers to a traditional CM Template $\mathcal{T}i$ focused on ad hoc security testing on the latest version of the build release ($T_{3,1}^s$).

$stage_4$ refers to stage Package. C_4 contains a single control $c_{4,1}$ executing dependency vulnerability checks on the additional packages ($T_{4,1}^s$), if any, required for application packaging.

$stage_5$ refers to stage Release. C_5 contains a single control $c_{5,1}$ verifying the correctness of the release procedure.

$stage_6$ refers to stage Configure. C_6 refers to two deployment CM Templates $\mathcal{T}_i^{dep}$ both having as target the deployment infrastructure ($T_{6,1}^s$): (i) $c_{6,1}$, defined in $\mathcal{T}_1^{dep}$, verifies

Fig. 4.3 An excerpt of the Development Certification Model Template $\mathcal{T}^{dev}$ for the DevOps process of the Assurance Storage service in Example 4.1.2

$stage_1$="Plan"
T_1^s={$T_{1,1}^s$="Solution design"}
C_1={c_1="Verify requirements document"}
Υ_1= =({$c_{1,1}$,T_1^s}, success)
$stage_2$="Create"
T_2^s= {$T_{2,1}^s$="Requirements document", $T_{2,2}^s$="Code repository"}
C_2= {$c_{2,1}$="Check requirements links", $c_{2,2}$="Code analysis"}
Υ_2= ({$c_{2,1}$,$T_{2,1}^s$}, success) ∧ ({$c_{2,2}$, $T_{2,2}^s$}, success)
$stage_3$="Verify"
T_3^s={ $T_{3,1}^s$="Latest release"}
C_3={$c_{3,1}$= $\mathcal{T}i$}
Υ_3= ({$c_{3,1}$, T_3^s}, success)
$stage_4$="Package"
T_4^s={$T_{4,1}^s$="Dependencies"}
C_4={$c_{4,1}$="Dependencies vulnerability checks"}
Υ_4= ({$c_{4,1}$ T_4^s}, success)
$stage_5$="Release"
T_5^s={$T_{5,1}^s$="Deployment environment"}
C_5={$c_{5,1}$="Release procedure checks"}
Υ_5=({$c_{5,1}$, T_5^s}, success)
$stage_6$="Configure"
T_6^s={$T_{6,1}^s$="Deployment Infrastructure"}
C_6={$c_{6,1}$=$\mathcal{T}_1^{dep}$, $c_{6.2}$=$\mathcal{T}_2^{dep}$}
Υ_6=({$c_{6,1}$, T_6^s}, success) ∨ ({$c_{6,2}$, T_6^s}, success)
$stage_7$="Monitor"
T_7^s={$T_{7,1}^s$="Application"}
C_7={$c_{7,1}$="Vulnerability scan", $c_{7,2}$="Pen test", $c_{7,3}$="Execution trace checks"}
Υ_7=({$c_{7,1}$, T_7^s}, success) ∧ ({$c_{7,2}$, T_7^s}, success) ∧ ({$c_{7,3}$, T_7^s}, success)

the compliance to the Center for Internet Security benchmark; (ii) $c_{6,2}$, defined in $\mathcal{T}_2^{dep}$, checks the absence of exploitable vulnerabilities by collecting metrics based on vulnerability scan and penetration testing. We note that the activities listed above are based on probes periodically executed on the infrastructure and the collected results are made available as metrics.

$stage_7$ refers to stage Monitor. C_7 contains three types of controls all having as target the system application instead of the infrastructure ($T_{7,1}^s$): (i) $c_{7,1}$ performs a vulnerability scan; (ii) c_2 performs a penetration testing; and (iii) c_3 monitors the executions of the target system when deployed in the operation environment [44].

Stages $stage_5$, $stage_6$, $stage_7$, which refers to DevOps stages Release, Configure, and Monitoring, respectively, need to be applied both in staging and operation environments.

We also note that, the Development Certification Model Template $\mathcal{T}^{dev}$ is a composite model where certification activities are independently specified for each DevOps stage. Differently from the other certification models, given the compositional nature of the template, a certification model instance is mapped on the DevOps process, where multiple instantiation functions λ_i generate the specific instances at each step. At a logical level, when the i-th DevOps stage is triggered, the corresponding λ_i is executed and the certification activities in $\mathcal{T}^{dev}$ for that stage are instantiated on the relevant system artifact T_i^s (e.g., code, package). We note that, if the verification of a given stage is described itself as a CM Template (e.g., stages verify and configure), the corresponding instantiation function λ_i generates and executes the specific CM Instance.

The result of the execution of the instantiation functions is an instance $\mathcal{I}^{dev}$ of the Development CM Template $\mathcal{T}^{dev}$, which drives real certification activities along the DevOps pipeline.

We recall that stage Plan in $\mathcal{T}^{dev}$ requires manual activities; therefore, the instantiation function executes manual checks. For performance reasons, the execution of instantiation functions λ can be pre-computed for each development round.

References

1. M. Anisetti, C. A. Ardagna, E. Damiani, A. Maña, G. Spanoudakis, L. Pino, and H. Koshutanski, "Security certification for the cloud: The cumulus approach," *Guide to Security Assurance for Cloud Computing*, pp. 111–137, 2015.
2. G. McGraw, "Software security," *IEEE Security & Privacy*, vol. 2, no. 2, 2004.
3. V. Lotz, "Cybersecurity Certification for Agile and Dynamic Software Systems - a Process-Based Approach," in *Proc. of EuroS&PW*, Genoa, Italy, September 2020.
4. M. Anisetti, C. A. Ardagna, and N. Bena, "Multi-dimensional certification of modern distributed systems," *IEEE Transactions on Services Computing*, 2022.
5. H. Teigeler, S. Lins, and A. Sunyaev, "Drivers vs. Inhibitors - What Clinches Continuous Service Certification Adoption by Cloud Service Providers?" in *Proc. of HICSS 2018*, Waikoloa, HI, USA, January 2018.
6. J. Lansing, A. Benlian, and A. Sunyaev, ""Unblackboxing" Decision Makers' Interpretations of IS Certifications in the Context of Cloud Service Certifications," *Journal of the Association for Information Systems*, vol. 19, 2018.
7. J. Prüfer, "Trusting privacy in the cloud," *Information Economics and Policy*, vol. 45, 2018.
8. Y. Bai, Y. Zhang, Y. Zhou, and L. T. Yang, "A non-functional property based service selection and service verification model," in *Proc. of UIC 2011*, Banff, Canada, September 2011.
9. K. M. Khan, A. Erradi, S. Alhazbi, and J. Han, "Security oriented service composition: A framework," in *Proc. of IIT 2012*, Al Ain, UAE, March 2012.
10. A.R.R. Souza et al., "Incorporating security requirements into service composition: From modelling to execution," in *Proc. of the ICSOC-ServiceWave 2009*, Stockholm, Sweden, November 2009.
11. L. Pino and G. Spanoudakis, "Finding secure compositions of software services: Towards a pattern based approach," in *Proc. of NTMS 2012*, Istanbul, Turkey, May 2012.
12. H. D. Vo, D. C. Phung, V. Q. Dung, and V.-H. Nguyen, "Securing data in composite web services," in *Proc. of KSE 2012*, Danang, Vietnam, August 2012.
13. M. Alrifai, T. Risse, and W. Nejdl, "A hybrid approach for efficient web service composition with end-to-end qos constraints," *ACM Transactions on the Web*, vol. 6, no. 2, pp. 1–31, 2012.
14. Y. Chen, J. Huang, C. Lin, and J. Hu, "A partial selection methodology for efficient qos-aware service composition," *IEEE Transactions on Services Computing*, vol. 8, no. 3, pp. 384–397, 2015.
15. S. Deng, H. Wu, D. Hu, and J. Zhao, "Service selection for composition with qos correlations," *IEEE Transactions on Services Computing*, vol. 9, no. 2, pp. 291–303, 2016.
16. H. Zheng, J. Yang, and W. Zhao, "Probabilistic qos aggregations for service composition," *ACM Transactions on the Web*, vol. 10, no. 2, pp. 12:1–12:36, May 2016.

17. M. Anisetti, C. Ardagna, and E. Damiani, "Security certification of composite services: A test-based approach," in *Proc. of ICWS 2013*, San Francisco, CA, USA, June–July 2013.

18. M. Anisetti, C. Ardagna, E. Damiani, and G. Polegri, "Test-based security certification of composite services," *ACM Transactions on the Web*, vol. 13, no. 1, pp. 1–43, 2018.

19. A. B. de Oliveira Dantas, F. H. de Carvalho Junior, and L. S. Barbosa, "A component-based framework for certification of components in a cloud of HPC services," *Science of Computer Programming*, vol. 191, 2020.

20. D. Redavid, L. Iannone, T. Payne, and G. Semeraro, "Owl-s atomic services composition with swrl rules," in *Proc. of ISMIS 2008*. Springer, 2008, pp. 605–611.

21. A. Alves et al., *Web Services Business Process Execution Language Version 2.0*, OASIS, April 2007, http://docs.oasis-open.org/wsbpel/2.0/OS/wsbpel-v2.0-OS.html.

22. C. A. Ardagna, M. Anisetti, B. Carminati, E. Damiani, E. Ferrari, and C. Rondanini, "A blockchain-based trustworthy certification process for composite services," in *Proc. of SCC 2020*. IEEE, 2020, pp. 422–429.

23. C. Redl, I. Breskovic, I. Brandic, and S. Dustdar, "Automatic SLA Matching and Provider Selection in Grid and Cloud Computing Markets," in *Proc. of ACM/IEEE Grid 2012*, Beijing, China, September 2012.

24. F. Jrad, J. Tao, A. Streit, R. Knapper, and C. Flath, "A utility/based approach for customised cloud service selection," *International Journal of Computational Science and Engineering*, vol. 10, 2015.

25. A. Taha, S. Manzoor, and N. Suri, "SLA-Based Service Selection for Multi-Cloud Environments," in *Proc. of IEEE EDGE 2017*, Honolulu, HI, USA, September 2017.

26. M. Anisetti, C. A. Ardagna, E. Damiani, and J. Maggesi, "Security certification-aware service discovery and selection," in *Proc. of SOCA 2012*. IEEE, 2012, pp. 1–8.

27. C. Shu, Z. Zhao, Y. Han, G. Min, and H. Duan, "Multi-User Offloading for Edge Computing Networks: A Dependency-Aware and Latency-Optimal Approach," *IEEE Internet of Things Journal*, vol. 7, no. 3, pp. 1678–1689, 2020.

28. M. Anisetti, C. A. Ardagna, N. Bena, and E. Damiani, "An assurance framework and process for hybrid systems," in *Communications in Computer and Information Science*, vol. 1484 CCIS, 2021, pp. 79–101.

29. N. Akhtar, A. Raza, V. Ishakian, and I. Matta, "COSE: Configuring Serverless Functions using Statistical Learning," in *Proc. of INFOCOM 2020*, 2020, pp. 129–138, iSSN: 2641-9874.

30. S. Nastic, T. Rausch, O. Scekic, S. Dustdar, M. Gusev, B. Koteska, M. Kostoska, B. Jakimovski, S. Ristov, and R. Prodan, "A Serverless Real-Time Data Analytics Platform for Edge Computing," *IEEE Internet Computing*, vol. 21, no. 4, pp. 64–71, 2017.

31. A. Das, A. Leaf, C. A. Varela, and S. Patterson, "Skedulix: Hybrid Cloud Scheduling for Cost-Efficient Execution of Serverless Applications," in *Proc. of CLOUD 2020*, 2020, pp. 609–618, iSSN: 2159-6190.

32. M. Anisetti, F. Berto, and R. Bondaruc, "Qos-aware deployment of service compositions in 5g-empowered edge-cloud continuum," in *IEEE CLOUD 2023*. IEEE, 2023.

33. M. Oulaaffart, R. Badonnel, and O. Festor, "Cmsec: A vulnerability prevention tool for supporting migrations in cloud composite services," in *Proc. of CloudNet 2022*. IEEE, 2022, pp. 267–271.

34. M. Anisetti, C. A. Ardagna, and E. Damiani, "Container-level security certification of services," in *Business System Management and Engineering: From Open Issues to Applications*. Springer, 2012, pp. 93–108.

35. M. Anisetti, C. A. Ardagna, F. Berto, and E. Damiani, "A security certification scheme for information-centric networks," *IEEE Transactions on Network and Service Management*, vol. 19, no. 3, pp. 2397–2408, 2022.

36. M. Anisetti, F. Berto, and M. Banzi, "Orchestration of data-intensive pipeline in 5g-enabled edge continuum," in *IEEE SERVICES 2022*. IEEE, 2022, pp. 2–10.

37. C. A. Ardagna, N. Bena, and R. M. de Pozuelo, "Bridging the gap between certification and software development," in *Proc. of ARES 2022*, Vienna, Austria, Aug. 2022.
38. J. H. Yahaya, A. Deraman, F. Baharom, and A. R. Hamdan, "Software certification from process and product perspectives," *International Journal of Computer Science and Network Security*, vol. 9, no. 3, pp. 222–231, 2009.
39. F. Baharom, J. Yahaya, A. Deraman, and A. R. Hamdan, "Spqf: software process quality factor," in *Proc. of ICEEI 2011*, Bandung, Indonesia, July 2011.
40. S. M. Darwish, "Software test quality rating: A paradigm shift in swarm computing for software certification," *Knowledge-Based Systems*, vol. 110, pp. 167–175, 2016.
41. M. Anisetti, C. A. Ardagna, F. Gaudenzi, and E. Damiani, "A continuous certification methodology for devops," in *Proc. of MEDES 2019*, 2019, pp. 205–212.
42. M. Anisetti, N. Bena, F. Berto, and G. Jeon, "A devsecops-based assurance process for big data analytics," in *Proc. of ICWS 2022*. IEEE, 2022, pp. 1–10.
43. M. Anisetti, C. A. Ardagna, E. Damiani, and F. Gaudenzi, "A semi-automatic and trustworthy scheme for continuous cloud service certification," *IEEE Transactions on Services Computing*, vol. 13, no. 1, 2020.
44. M. Anisetti, C. A. Ardagna, E. Damiani, N. El Ioini, and F. Gaudenzi, "Modeling time, probability, and configuration constraints for continuous cloud service certification," *Computers & Security*, vol. 72, pp. 234–254, 2018.

Beyond Cloud Service Certification

5

Cloud computing has been a driving force for many technological innovations and transformations in various domains and industries. It offers scalable and cost-effective storage and processing capabilities suitable for handling large volumes of structured and unstructured data. Could computing has enabled organizations to leverage data-driven insights and decision making, as well as to create new products and services based on data analysis. Today, cloud computing offers access to powerful hardware and software resources for developing and deploying Artificial Intelligence (AI) and Machine Learning (ML) applications that perform different tasks such as natural language processing, computer vision, speech recognition, and recommendation systems. From the software architecture and the software development points of view, however, these applications are nothing but distributed systems, deployed on the cloud-edge continuum enabled by the 5G mobile network. The sheer amount, size and complexity of these distributed systems based on AI and ML has fundamentally changed their behavior [1]. Certification schemes have to keep the pace of this continuous system evolution, which opens the door to multiple challenges. In this chapter, we analyse the need of redesigning certification schemes to address the requirements introduced by modern distributed systems underlying ML and AI applications.

5.1 Certification of Cloud-Edge Distributed Systems

Modern systems are moving from centralized architectures where computations are done at the core of the network, to decentralized architectures where important parts of the computations are executed at the edge of the network, near the data collection points. This environment is often referred to as the *cloud-edge continuum* [2], where complex compu-

© The Author(s), under exclusive license to Springer Nature Switzerland AG 2025
M. Anisetti et al., *A Journey into Security Certification*, Synthesis Lectures on Information Security, Privacy, and Trust, https://doi.org/10.1007/978-3-031-59724-4_5

tation pipelines are deployed on the continuum and connected to the data sources that are often localized in the Internet of Things.

Internet of Things (IoT) can be defined as *"the networked interconnection of everyday objects, equipped with ubiquitous intelligence"* [3]. It is composed of physical devices from minuscule sensors to big components that are connected by fast wired and mobile networks, and contributes to the implementation of smart services delivered at the edge. IoT is a fundamental building block of the cloud-edge continuum connecting billions of devices sensing the physical environment through high-throughput and ultra-low latency (5G) communications. This architecture pushed forward a data-driven ecosystem, where a huge amount of data are created, collected, consumed, and stored, and cannot be effectively handled using traditional cloud infrastructures.[1] Edge nodes are then used to pre-process data at the edge and forward them to the cloud where cloud-based applications are built on top of processing services.

This new digital ecosystem comes with important advantages in terms of applications and added value for its users, making their world smarter and simpler. However, the price we pay for such advantages is an increasing difficulty in managing non-functional aspects including security, privacy and trustworthiness, which are must-have requirements in scenarios where the users are yet another system component, and are surrounded and interact with a multitude of systems/sensors.

5.1.1   Challenges

The redesign of certification schemes for the cloud (Sect. 3.3), is driven by a new set of challenges as follows.

- *Verification of heterogeneous and complex systems with unknown boundaries.* Modern distributed systems mix heterogeneous technologies in a multi-layer infrastructure, where a multitude of smart devices are connected through fast and reliable networks (e.g., 5G network). The certification of these systems carries multiple challenges mostly connected to the variability of the environment and the intrinsic lack of trustworthiness of "bring you own" (BYO) smart devices, which are often under the control of potentially unreliabke end-users. More in detail, the reference environment quickly evolves over time in terms of participating entities and topology, resulting in an environment with unknown boundaries and diverse technologies. Also, smart devices are usually resource-constrained, physically accessible, and protected with lower standards than the traditional cloud computing nodes.

---

[1] Statista https://www.statista.com/statistics/871513/worldwide-data-created/ claimed that the total amount of data created, captured, copied, and consumed globally is expected to grow to more than 180 zettabytes by 2025, a number that is also shared by IDC (https://www.forbes.com/sites/tomcoughlin/2018/11/27/175-zettabytes-by-2025/?sh=2404c2385459) expecting to grow to 175 zettabytes by 2025.

This breaks two of the assumptions at the basis of traditional cloud certification schemes: (i) the ability to precisely define the boundaries of the target of certification; (ii) the trustworthiness of the target of certification as discussed in the following challenge.

- *Management of untrustworthy data and providers.* During multi-factor certification of composite systems, information about the system components, the network infrastructure, and the execution environment is collected and used to compute of verify the certificates. Some traditional certification schemes made the assumption of being able to collect trustworthy data from trustworthy/known sources. Modern systems, especially the ones including smart devices, cannot guarantee the above assumptions, since data are continuously collected from a multitude of devices, which are intrinsically unreliable and under the control of many un-trusted users and service providers.

- *Zero-trust approach* Certification schemes must depart from any assumption on data or provider trustworthiness and adopt instead a *zero-trust* approach. Zero-trust approaches [4] are based on the notion that no participant to any process is trusted a priori. Thus, all data sources and computing services used in the certification process must be treated as resources that must be secured. No trust should be automatically granted to test and certification tools, and no actors requesting access (e.g., to collect evidence) should be trusted by default. Zero-trust implementation of the certification process requires specific security controls: for instance, access to evidence must be granted per session only (i.e., not granted statically but continuously re-evaluated), and least privilege principle should be carefully observed, avoiding assigning root privileges to certification tools.

- *Certification of resource-constrained devices.* Modern certification schemes must be designed to have a limited impact on the target of certification. The latter, being composed of many resource-constrained devices, cannot be certified in deep with the risk of exhausting system resources. Evidence collection techniques (e.g., testing, monitoring) must balance between the completeness of retrieved evidence and the impact on the target components. Full certification can come at the price of being harmful to the system under certification, and must then be relaxed to limit the impact on its functioning.

- *Behavioral testing vs component testing.* As discussed in Chap. 3 traditional certification schemes are based on the detailed modeling of the target of certification and its components, which are monitored and tested both independently and in composition. Modern certification schemes must depart from this assumption and consider behavioral testing as a possible approach to certification. Behavioral testing consists of the modeling of the expected system behavior that is then matched against the observed system behavior [5].

5.1.2   Machine Learning for Cloud-Edge Certification

A certification scheme for the cloud-edge continuum must address three main requirements addressing the challenges in Sect. 5.1.1: (i) a novel certification model capturing the behavior of a complex system in the cloud-edge continuum, (ii) a distributed evidence collection process with minimum overhead on the target system and its resource-constrained devices, (iii) a transparent and continuous certification process that updates its model and corresponding certificates according to system changes.

As discussed in Sect. 3.3, traditional certification models are composed of three main parts specifying the property to be certified, the target of certification, the evidence collection model. The certification model represents the starting point of a novel certification process for the cloud-edge continuum. In particular, upon statically certifying a cloud-edge system in a controlled environment, a machine-learning based certification process should start to keep the validity of the certificate in operation and across system changes. The inherent format of a cloud-edge environment, with no fixed topology and resource-constrained devices, in fact, requires new ways of modeling non-functional properties, as well as new ways of modeling the target of certification and collecting evidence to address the challenges in Sect. 5.1.1. Non-functional properties must be redesigned to model requirements in multiple dimensions of interest (e.g., artifacts, development, evaluation) [6] and adapt to the system changes. Machine-learning techniques can be used to model the behavior of the system under certification and drive evidence collection, managing the entire certificate life cycle. In particular, a continuous certification process can be defined, where the target of certification and the evidence collection model flows in a single behavioral component as follows:

1. Upon certifying the target of certification in a controlled environment, a model of its behavior is built using machine learning, with no requirements on the topology of the target or static assumptions on its composition. The behavioral model is built before the system is put into operation and becomes the cornerstone for certificate maintenance and continuous verification.
2. The behavioral model is then continuously checked against the model derived by the system in operation. As soon as a misalignment between the original behavioral model and the current behavioral model is detected (anomaly detection), an adaptation process is raised.
3. The adaptation process retrieves the differences between the two models and executes corrective actions that result in the (i) confirmation of the original certificate, (ii) refinement of the original certificate in a new one modeling the current status of the system, (iii) revocation of the certification and triggering of re-certification from scratch.

In summary, service certification consists of the release of a certificate in a controlled environment using traditional certification schemes, followed by the modeling of the expected behavior of the system as a whole and the continuous monitoring of certificate validity through an anomaly detector spotting any divergence from the expected behavior.

5.2   Certification of Machine Learning-Based Systems

Traditional software systems mostly implemented deterministic algorithms, while modern systems are increasingly managed by Machine Learning (ML) models, relying on distributed protocols for their training and execution. ML models are at the core of modern systems and drive their behavior, that is, they reason on data to calculate a solution to individual instances of a problem [7].

ML models are deeply changing distributed systems' design and development, as well as their monitoring and verification. They are often treated as black boxes that make unpredictable, automated decisions. Lack of transparency can make it difficult to quantify safety and security risks for the final users. Lack of transparency refers to the insufficient or unclear disclosure of information, such as the source, scope, and validity of data, the methods and assumptions of analysis, or the criteria and standards of evaluation. Lack of transparency can affect the trust in a system in several ways:

- *False sense of security*: If the system's users do not have access to the relevant and reliable information about the models, they may make decisions based on incomplete, inaccurate, or outdated data. For example, if an AI-ML system does not explain how it generates its outputs, its users may not be aware of the potential errors, biases, or limitations of the system.
- *User Disengagement*: If the system's users do not have a clear understanding of how and why decisions are made, they may feel excluded, ignored, or manipulated by the decision makers. They may stop using the system altogether, or they may question the fairness, or accountability of the outputs.
- *Supplier Abuse*: If the users cannot monitor or verify the actions or behaviors of the system involved, they may be exposed to the risk of abuse or fraud by suppliers who have hidden agendas, or conflicts of interest.

Indeed, transparency is important for ensuring the safety and security of the final users, enabling to monitor and verify the actions and behaviors of all actors involved, and to hold them accountable and responsible for the consequences and impacts of their actions and behaviors.

This requirement makes verification of ML-based system non-functional properties a major challenge by itself, and requires yet another call for new certification schemes that accomplish the peculiarities of machine learning. Lack of trustworthiness is in fact one of the main factors hampering ML-based system adoption in critical domains. In support of this view, Yoshioka and Ishikawa [8] conducted a study involving several industry experts, which claimed that the assessment of non-functional requirements is among the most complex stages of a ML-based system development. ML models are difficult to explain and monitor, thus impeding the adoption of assurance approaches including certification [9].

In this context, new guidelines and regulations are defined/under definition concerning ML/AI trustworthiness, ethics, and risk. For instance, the high-level Expert Group on AI of the European Union presented the *Ethics Guidelines for Trustworthy Artificial Intelligence*[2] claiming that: "*As it cannot be expected that everyone is able to fully understand the workings and effects of AI systems, consideration can be given to organisations that can attest to the broader public that an AI system is transparent, accountable and fair. These certifications would apply standards developed for different application domains and AI techniques, appropriately aligned with the industrial and societal standards of different contexts.*" In 2021, the European commission then released the AI-act[3] claiming compliance and risk management as fundamental requirements for high-risk AI systems. This work agrees on the need of solutions for evaluating the non-functional shape of ML/AI systems, including certification, the focus of this book.

5.2.1 Challenges

Modern distributed systems based on ML and artificial intelligence pose strict requirements on the redesign of current certification schemes. In particular, new schemes must be defined to support the certification of ML models, as follows.

- *Non-functional properties definition.* The non-functional verification of ML-based systems make current definition of non-functional properties (see Sect. 3.3) unusable. Attribute-based properties do not fit non-deterministic systems whose behavior is not known a priori. They must be replaced with properties that provide either a statistical, probabilistic or behavioral-based definition.
- *ML models are treated as black boxes.* The certification of ML based-service requires the ability to verify the ML black box. The emerging domain of ML verification uses behavioral monitoring techniques based on statistical testing to dynamically check properties (besides accuracy and fairness) of black-box ML models in operation. Monitors can be used to generate certificates for software applications, and composed to obtain dynamic virtual certificates for ensemble ML models. This however requires a proper definition of the entire certification chain, which must depart from the verification of deterministic systems based on traditional testing and consider the variability in behavior introduced by training data, configuration, and process.
- *Adaptive behavior.* ML-based systems are often unable to adapt their behavior to changing conditions, especially when non-functional aspects are considered. Degraded behavior is mostly considered in terms of accuracy and precision of the ML-based system, while no adaptation is possible when non-functional properties like fairness are considered.

---

[2] https://digital-strategy.ec.europa.eu/en/library/ethics-guidelines-trustworthy-ai.
[3] https://artificialintelligenceact.eu/the-act/, https://digital-strategy.ec.europa.eu/en/policies/regulatory-framework-ai.

Also, it is usually not possible to adapt the ML-based behavior to contextual information as for instance the battery remaining on the specific devices executing the ML model. New certification scheme must be extended to support adaptive behavior of ML-based systems.

- *Statistical testing.* As discussed in Sect. 5.1.1, traditional certification schemes are based on the detailed modeling of the target of certification and its components, which are verified both independently and in composition. This approach does not fit well ML-based systems, which require statistical testing and monitoring of ML models' behavior at inference time to efficiently check that the desired behavioral properties are achieved [10].
- *Evaluation of ML pipelines.* Similarly to traditional cloud-based certification, current systems are often built as a composite service, often referred to as machine learning pipelines. Certification of ML pipelines is inherently multi-layer and multidimensional. First of all, each service in the pipeline must be certified within its deployment infrastructure. Afterwards, the certified services in the machine learning pipeline must be certified as a composite service verifying the orchestration of their functionalities.

5.2.2 Certification Modeling

To address the above challenges on ML certification, it is necessary to reshape traditional certification schemes by revisiting the three main building blocks of the certification model in Sect. 3.3: non-functional properties, target of certification, evidence. A first approach has been presented in [11].

Non-functional Properties (p). ML certification requires the definition of ML specific non-functional properties that redesign the way in which properties are defined and evaluate. Statistical and behavioral-based properties are needed to model the peculiarities of ML models. Some taxonomies have been already described (e.g., [10] and[4]) preliminary defining some properties as follows:

- *Transparency*: the feasibility of ex-post interpretation. It can further be specialized in *Explainability* [12] and *Interpretability* [13].
- *Fairness*: the absence of discrimination, that is, prejudice against an individual or a group based on the value of specific features [14].
- *Legality*: the compliance to laws and regulation, including privacy, quality, fundamental human rights, fairness, civil and criminal liability [15].
- *Maintainability*: the ability of going through complete or partial model retraining [16].
- *Modularity*: the ability to manage large or highly complex systems [17].

---

[4] https://www.enisa.europa.eu/publications/artificial-intelligence-cybersecurity-challenges?v2=1.

- *Performance*: the performance of a computation in terms of accuracy, precision/recall, and Area Under the ROC curve.
- *Privacy*: the requirement of deleting information about the training set and those inferred from the models' output.
- *Reliability*: the ability to operate with no failures and guarantee a level of performance when operating in normal conditions [18].
- *Safety*: "*the expectation that a system does not, under defined conditions, lead to a state in which human life, health, property, or the environment is endangered*" [19].
- *Scalability*: the ability to adapt to environmental changes and guarantee a sufficient level of performance.
- *Robustness*: the robustness against targeted or untargeted attacks. It includes the ability to protect data, models, hyperparameters, weights and coefficients used by models, and proprietary algorithms.
- *Usability*: the effort required by users to learn the software, prepare input data and interpret the results.

Target of Certification (*ToC*). It is natively multi-layer and includes three main factors to be considered: (i) *data* (*Df*) including the description and metadata of the training, test, and validation sets. For instance, it can evaluate the granularity, and sparsity, to name but a few; (ii) *process* (*Pf*) describing the details of the training process such as the preprocessing applied to data, the type of learning that is implemented, and any other requirement on the model training; (iii) *model* (*Mf*) describing the model and its behavior.

Evidence Collection Model ($\mathcal{E}^*$). It specifies all activities needed to collect evidence on a specific factor of the target of certification. It ranges from the verification of the correct behavior of the model to the analysis of the possible data leak caused by observing the model in operation.

Our Certification for ML is therefore made of three independent certification models $[\mathcal{CM}^{Pf}, \mathcal{CM}^{Df}, \mathcal{CM}^{Mf}]$, one for each factor, implementing their specific certification process.

The *data factor certification model* $\mathcal{CM}^{Df} = \langle p^{Df}, ToC^{Df}, \mathcal{E}^{Df} \rangle$ evaluates the training validation and testing data and their impact on the ML model under certification. The direct relation between the quality of the training set and the performance of the trained ML model is a well-known problem. The increasing complexity and size of training sets make proper fine/tuning difficult in practice. $\mathcal{CM}^{Df}$ aims to address this gap and includes a property p^{Df} specific for data, a target ToC^{Df} modeling the data used for training/validation, and an evidence collection model $\mathcal{E}^{Df}$ specifying the procedure for collecting evidence on the ToC, including evidence on data balancing, granularity, and fine/tuning.

The *process factor certification model* $\mathcal{CM}^{Pf} = \langle p^{Pf}, ToC^{Pf}, \mathcal{E}^{Pf} \rangle$ evaluates the impact of the training process on the ML model under certification. For instance, training an ensemble instead of a monolithic model can have an impact on property robustness [20, 21], increasing the strength of the training process and compensating for weaknesses of the training set. $\mathcal{CM}^{Df}$ includes a property p^{Pf} specific for the training process, a target ToC^{Pf} modeling the training process, and an evidence collection model $\mathcal{E}^{Pf}$ including the procedure for collecting evidence on how the training process is designed and executed.

The *model factor certification model* $\mathcal{CM}^{Mf} = \langle p^{Mf}, ToC^{Mf}, \mathcal{E}^{Mf} \rangle$ evaluates the software artifact implementing the ML model which is a crucial and strongly intertwined with the property to be verified. For instance, how to model property fairness [22, 23] impacts on how to evaluate the ML model with respect to fairness. $\mathcal{CM}^{Mf}$ includes a property p^{Mf} specific for ML model, a target ToC^{Mf} modeling the ML model artifact (e.g., its structure, weights constraints), and an evidence collection model $\mathcal{E}^{Mf}$ including the procedure for collecting evidence on the ML model behavior, such as functions for exercising the model itself.

5.2.3 Executing Certification of ML

We now present three practical scenarios to demonstrate our multi-factor approach, each considering a different non/functional property. For simplicity, each property is certified according to one factor only. More complex scenarios might require more than one factor at a time. We describe (i) process factor (Pf) considering property *robustness*, (ii) data factor (Df) considering property *absence of targeted poisoning*, and (iii) model factor (Mf) considering property *fairness*. Our examples are summarized in Table 5.1.

Process factor and property robustness. We consider an ensemble of $N = 9$ random forests for email spam detection to be certified for property robustness. A certification model $\mathcal{CM}^{Pf} = \langle p^{Pf}, ToC^{Pf}, \mathcal{E}^{Pf} \rangle$ is defined as follows.

Property p^{Pf} is defined as the difference Δ between the accuracy of the target ML model trained on a clean dataset and the one on the poisoned dataset [21]. The target ML model is considered robust if Δ is below a threshold.

Target ToC^{Pf} is a training process that (i) splits the training set into random, non-/overlapping partitions according to a hashing function and modulus, (ii) trains each random forest independently. We used the popular Spambase dataset.[5]

Evidence collection model $\mathcal{E}^{Pf}$ first splits the dataset in training set and held/out test set, and poisons the training set by randomly flipping some labels. It then trains one ensemble on the clean dataset and one ensemble on the poisoned dataset. The accuracy of the two ensembles is calculated according to the final predictions on the test set. Each prediction is retrieved with majority voting among the random forests in the ensemble. Δ is then calculated and represents the collected evidence.

In our scenario, we retrieved $\Delta = 0.3$ modeling the larger robustness of an ensemble approach against poisoning [20, 21]. Assuming an upper threshold $\Delta = 5$; the certificate for property robustness and factor process Pf can be awarded.

---

[5] https://archive.ics.uci.edu/ml/datasets/Spambase, with 3626 data points and 57 features after pre-processing, referred to spam and non/spam emails.

Table 5.1 Summary of our three evaluation scenarios

f	p	ToC	$\mathcal{E}$	Outcome
Pf	Robustness	Ensemble of random forest	Difference in accuracy between poisoned and non/poisoned model evidence: $\Delta = 0.3$	Awarded
Df	Absence of targeted poisoning	Training set	Check on the presence of suspicious data points evidence: suspicious data points= $\emptyset$	Awarded
Mf	Fairness	Nearest neighbor	Variance of predictions over combinations of protected attributes evidence: $\sigma^2 = 220656800$	Not awarded

Data factor and property absence of poisoned data. We consider a neural network model to be trained for object recognition. A certification model $\langle p^{Df}, ToC^{Df}, \mathcal{E}^{Df} \rangle$ is defined as follows.

Property p^{Df} refers to the lack of any suspicious data points in the training set. Suspicious data points are likely to be maliciously/crafted data points pushing ML models in misclassifying specific data points at inference time.

Target ToC^{Df} is the training set.

Evidence collection model $\mathcal{E}^{Df}$ adapts a poisoning removal technique such as [24], a kNN/based approach to remove suspicious data points, to identify the suspicious data points in the dataset.

In our scenario, we retrieved no suspicious data points, and a certificate for property robustness and factor data Df can be awarded.

Model factor and property fairness. We consider a nearest neighbor ML model to be trained for bail estimation, that is, to determine the amount of bail for people awaiting a trial. A certification model $\mathcal{CM}^{Mf} = \langle p^{Mf}, ToC^{Mf}, \mathcal{E}^{Mf} \rangle$ is defined as follows.

Property p^{Mf} refers to the absence of bias and discrimination in a ML model. We note that modeling the concept of fairness is rather complex; more than 20 different (and sometimes conflicting [22]) definitions of fairness are available in literature [23]. We consider a simple notion of fairness based on the identification of sensitive features: the target ML model is considered fair if the variance σ^2 of the predicted output is below a given threshold when varying the protected features (Mf). We used the dataset of the Connecticut State Department of Correction.[6]

Target ToC^{Df} is the trained nearest neighbor model.

Evidence collection model $\mathcal{E}$ first trains the nearest neighbor model on a portion of the whole dataset. It then generates, for each data point in the test set, a set of artificial data points covering all the possible combinations of protected features (race and gender) using Monte Carlo simulation. The collected evidence is expressed as the variance σ^2 of the ML model predictions over the generated data points. We repeated the latter process two times, while varying the data points used at each run [10].

In our scenario, we retrieved fairness $\sigma^2 = 220656800$, which is above the threshold (200). The certificate for property fairness and Mf factor cannot therefore be awarded.

References

1. D. Lindsay, S. S. Gill, D. Smirnova, and P. Garraghan, "The evolution of distributed computing systems: from fundamental to new frontiers," *Computing*, vol. 103, no. 8, pp. 1859–1878, 2021.
2. K. Fu, W. Zhang, Q. Chen, D. Zeng, and M. Guo, "Adaptive resource efficient microservice deployment in cloud-edge continuum," *IEEE Transactions on Parallel and Distributed Systems*, vol. 33, no. 8, pp. 1825–1840, 2021.
3. F. Xia, L. Yang, L. Wang, and A. Vinel, "Internet of things," *International Journal of Communication Systems*, vol. 25, no. 9, pp. 1101–1102, 2012.
4. C. Buck, C. Olenberger, A. Schweizer, F. Völter, and T. Eymann, "Never trust, always verify: A multivocal literature review on current knowledge and research gaps of zero-trust," *Computers & Security*, vol. 110, p. 102436, 2021.
5. C. A. Ardagna and N. Bena, "Non-functional certification of modern distributed systems: A research manifesto," in *Proc. of SSE 2023*, Chicago, IL, USA, July 2023.
6. M. Anisetti, C. A. Ardagna, and N. Bena, "Software test quality rating: A paradigm shift in swarm computing for software certification," *IEEE Transactions on Services Computing*, 2022.
7. E. Damiani and C. Ardagna, "Certified machine-learning models," in *Proc. of SOFSEM 2020*, Limassol, Cyprus, January 2020.
8. F. Yoshioka and I. N., "How do engineers perceive difficulties in engineering of machine-learning systems? - Questionnaire survey," in *Joint Intl. Workshop on Conducting Empirical Studies in Industry and Intl. Workshop on Software Engineering Research and Industrial Practice*, 2019.
9. C. Ardagna, R. Asal, E. Damiani, and Q. Vu, "From Security to Assurance in the Cloud: A Survey," *ACM Computing Surveys*, vol. 48, no. 1, 2015.
10. M. Anisetti, C. A. Ardagna, E. Damiani, and P. G. Panero, "A methodology for non-functional property evaluation of machine learning models," in *Proc. of MEDES 2020*, ser. MEDES '20. New York, NY, USA: Association for Computing Machinery, 2020, p. 38-45. [Online]. Available: https://doi.org/10.1145/3415958.3433101
11. M. Anisetti, C. A. Ardagna, N. Bena, and E. Damiani, "Rethinking certification for trustworthy machine-learning-based applications," *IEEE Internet Computing*, vol. 27, no. 6, 2023.

---

[6] https://data.ct.gov/Public-Safety/Accused-Pre-Trial-Inmates-in-Correctional-Faciliti/b674-jy6w, downloaded Feb. 21st, 2020, containing 4,182,246 data points and 8 features referred to the individuals, such as race, gender, offence.

12. J. Winkler and A. Vogelsang, "What does my classifier learn? A visual approach to understanding natural language text classifiers," in *Proc. of NLDB 2017*, 2017.
13. R. Gall, "Machine Learning Explainability vs Interpretability: Two concepts that could help restore trust in AI," 2018. [Online]. Available: https://www.kdnuggets.com/2018/12/machine-learning-explainability-interpretability-ai.html
14. L. Hendricks, K. Burns, K. Saenko, T. Darrell, and A. Rohrbach, "Women also Snowboard: Overcoming Bias in Captioning Models," in *Computer Vision – ECCV*. Springer, 2018.
15. Global Legal Research Directorate, "Regulation of Artificial Intelligence in Selected Jurisdictions," The Law Library of Congress, Tech. Rep. January, 2019. [Online]. Available: https://www.loc.gov/law/help/artificial-intelligence/index.php
16. M. Schmitz, "Why your Models need Maintenance," 2017. [Online]. Available: https://towardsdatascience.com/why-your-models-need-maintenance-faff545b38a2
17. C. E. Perez, "Modularity," pp. 1–15, 2018. [Online]. Available: https://www.deeplearningpatterns.com/doku.php?id=modularity
18. D. Mairiza, D. Zowghi, and N. Nurmuliani, "An investigation into the notion of non-functional requirements," *Proc. of SAC 2010*, pp. 311–317, 2010.
19. ISO/IEC and IEEE, "ISO/IEC/IEEE 24765:2010 - Systems and software engineering – Vocabulary," *Iso/Iec Ieee*, vol. 2010, p. 410, 2010. [Online]. Available: http://www.iso.org/iso/catalogue_detail.htm?csnumber=50518
20. A. Levine and S. Feizi, "Deep Partition Aggregation: Provable Defenses against General Poisoning Attacks," in *Proc. of ICLR 2021*, Vienna, Austria, May 2021.
21. M. Anisetti, C. A. Ardagna, A. Balestrucci, N. Bena, E. Damiani, and C. Y. Yeun, "On the robustness of random forest against untargeted data poisoning: An ensemble-based approach," *IEEE Transactions on Sustainable Computing*, pp. 1–15, 2023.
22. J. Kleinberg, S. Mullainathan, and M. Raghavan, "Inherent Trade-Offs in the Fair Determination of Risk Scores," *arXiv preprint* arXiv:1609.05807, 2016.
23. D. Pessach and E. Shmueli, "A review on fairness in machine learning," *ACM CSUR*, vol. 55, no. 3, 2022.
24. N. Peri, N. Gupta, W. R. Huang, L. Fowl, C. Zhu, S. Feizi, T. Goldstein, and J. P. Dickerson, "Deep k-NN Defense Against Clean-Label Data Poisoning Attacks," in *Proc. of ECCV 2020*, Glasgow, UK, August 2020.

Conclusions and Open Issues

6

Since its early introduction in the Eighties, certification has been a primary way to document the verification of systems behavior. Certification schemes and corresponding processes have been defined for software systems, first, and the adapted to service-based and cloud-based systems. Finally their scope has included ML-based systems. Certification schemes implement a procedure by which a trusted third party gives written assurance (evidence) that a product, process or service hold some non-functional properties and behaves correctly.

In this book, we retraced the history of IT systems' certification and concentrated on security certification of modern distributed systems. We discussed the evolution of service certification focusing on the verification of cloud services and the need of a solution that is trustworthy, continuous, incremental, and adaptive. We then shifted our attention to modern distributed systems, presenting a reference certification scheme that addresses the entire system architecture, from the certification of service compositions to the certification of the target deployment environment, through the certification of the entire development process. Finally, we presented two domains currently driving research on certification, namely, certification of cloud-edge distributed systems and certification of machine learning-based systems. In both these domains, software certification should be automated to some extent to match the system variants at execution time. Our reference dynamic certification scheme can be used to automate the verification and validation of software, running unit tests, integration tests, regression tests, and security tests as part of an integrated development-deployment cycle. Certification becomes integral part of Continuous Integration and Continuous Delivery (CI/CD) methods for automating the building, testing, and deployment stages. As we have seen, CI/CD can be used to automate the delivery and certification of software, such as compiling the code, running the tests, deploying the software, or monitoring the performance. However, software certification process automation also has some challenges, such as:

- Requirements diversity and variability: While dynamic certification can handle changes in the target of certification structure and components, software certification may involve variable requirements, standards, or criteria. Some requirements, like cultural acceptability or fairness of software products may still require human judgment, creativity, or expertise to assess.
- Accuracy and reliability: Software certification may depend on the accuracy and reliability of the software tools and frameworks used to automate the process. Of course, one could think of "certifying the certifier's tools": but some root-of-trust schemes are needed to avoid unnecessary recursion.
- Cost and investments: Automating software certification may require significant investment to acquire, develop, and update the software tools and frameworks used to automate the process.

Therefore, the automation of software certification should not be seen as a "turn-key" solution. Indeed, automating the software certification process requires careful planning, design and implementation to ensure that it meets the specific needs and goals of the software products, processes, or services being certified.

6.1   Research Directions

Being closely related to IT systems evolution, research on system certification is continuously evolving, facing new issues and challenges. A lot is still to be done to achieve a solution that addresses the peculiarities of modern AI-based systems at best. Several open research questions need to considered [1]. We briefly discuss them below.

Certification model and certificate definition. Modern distributed systems are changing certification schemes at their roots, requiring a complete revision of all their building blocks. First of all, non-functional properties should be evaluated over time according to system changes, possibly supporting probabilistic and fuzzy evaluation. Then, the notion of certification target should be widened to include increasingly complex and opaque systems based on ML. The evidence collection model should support behavioral evidence beyond traditional test- and monitoring-based evidence. The latter should be at the basis of a continuous certification process, which considers and tracks the evolution of the observed behavior. Finally, the certificate must become a function of time and go beyond the simple awarded/not awarded scenario, and consider metrics supporting the magnitude entailed by the certificate itself.

ML for certification. Automation of certification must be a cornerstone of certification schemes. ML should therefore be a must-have to boost automation in generating certification model building blocks and in executing continuous certification processes. ML should also be used to model the expected behavior of a ToC at the basis of continuous certificate validity and enhanced certificate life cycle.

Certification of ML. As we already discussed in Sect. 5.2, certification of ML (and applications/systems build on it) is still in its infancy. Certification processes and all their building block should be re-though to address ML peculiarities. This research direction has an impact on all research directions in this section, due to the fact that ML is permeating every aspect of modern distributed systems and certification schemes.

Design and development meet certification. The full potential of certification can be achieved if the certification life cycle and the distributed system life cycle are merged in a single one. In this context, DevOps should be extended to DevCertOps, meaning that novel service engineering processes would be unified with the certification process. In other words, certification should become a source of requirements driving the development processes and paving the way to the implementation of certification-ready systems. This research direction should contribute to limit two of the main issues of certification monetary costs and time overhead. At a later stage (Ops), certification should contribute to the definition of systems with stable behavior, where certificate-based adaptation facilitates maintaining system non-functional properties across time and systems changes.

Block-chain-based certification. Distributed system evolution is a fast and inexorable process. Certification should keep the pace with this evolution and define new approaches addressing the need of edge-cloud continuum, 5G networks, and even satellite-based communications. Distributed ledgers are a promising technique for storing trustworthy use cases [2]. The ledger stores the input data used in test cases to simulate different scenarios and test the behavior of the system to be certified.

References

1. C. A. Ardagna and N. Bena, "Non-functional certification of modern distributed systems: A research manifesto," in *Proc. of SSE 2023*, Chicago, IL, USA, 2023.
2. A. Al Zaabi, C. Y. Yeun, and E. Damiani, "Trusting testcases using blockchain-based repository approach," *Symmetry*, vol. 13, no. 11, p. 2024, 2021.